Laboratories

Briefing and Design Guides

Series editor Maritz Vandenberg

The *Briefing and Design Guide* series is a growing library of practical manuals for building designers and the clients who brief them. With a deliberate focus on nuts-and-bolts design rather than glamorous photographs or esoteric theory, these books are planned for maximum usefulness to architects, engineers and clients. They should also offer an accessible source of reference to students.

Laboratories
Walter Hain

E & FN SPON
An Imprint of Chapman & Hall

London · Glasgow · Weinheim · New York · Tokyo · Melbourne · Madras

Published by E & FN Spon, an imprint of Chapman & Hall, 2–6 Boundary Row, London SE1 8HN, UK

Chapman & Hall, 2–6 Boundary Row, London SE1 8HN, UK

Blackie Academic & Professional, Wester Cleddens Road, Bishopbriggs, Glasgow G64 2NZ, UK

Chapman & Hall GmbH, Pappelallee 3, 69469 Weinheim, Germany

Chapman & Hall USA, One Penn Plaza, 41st Floor, New York NY 10119, USA

Chapman & Hall Japan, ITP-Japan, Kyowa Building, 3F, 2-2-1 Hirakawacho, Chiyoda-ku, Tokyo 102, Japan

Chapman & Hall Australia Thomas Nelson Australia, 102 Dodds Street, South Melbourne, Victoria 3205, Australia

Chapman & Hall India, R. Seshadri, 32 Second Main Road, CIT East, Madras 600 035, India

First edition 1995

© 1995 Walter Hain

This book was commissioned by Maritz Vandenberg for E & FN Spon

Cover photograph: Writtle College, teaching laboratory
Photographer: Matthew Weinreb
Lab designer: Llewelyn-Davies Architects

Typeset in 9/12 Univers Med by Florencetype Ltd, Stoodleigh, Devon

Printed in Great Britain by The Alden Press, Oxford

ISBN 0 419 19480 0

Apart from any fair dealing for the purposes of research or private study, or criticism or review, as permitted under the UK Copyright Designs and Patents Act, 1988, this publication may not be reproduced, stored, or transmitted, in any form or by any means, without the prior permission in writing of the publishers, or in the case of reprographic reproduction only in accordance with the terms of the licences issued by the Copyright Licensing Agency in the UK, or in accordance with the terms of licences issued by the appropriate Reproduction Rights Organization outside the UK. Enquiries concerning reproduction outside the terms stated here should be sent to the publishers at the London address printed on this page.

The publisher makes no representation, express or implied, with regard to the accuracy of the information contained in this book and cannot accept any legal responsibility or liability for any errors or omissions that may be made.

A catalogue record for this book is available from the British Library

∞ Printed on acid-free text paper, manufactured in accordance with ANSI/NISO Z39.48-1992 (Permanence of Paper).

Contents

vi	Acknowledgements
1	**1 Introduction**
1	1.1 Aims
2	1.2 The laboratory as a building type
5	1.3 Regulations and standards
6	1.4 The brief
9	1.5 Initial procedures in planning laboratories
13	**2 Laboratory spaces**
13	2.1 The laboratory space
22	2.2 Ancillary spaces
31	2.3 Storage areas
32	2.4 Workshops
32	2.5 Animal areas
35	2.6 Circulation
37	**3 Fitting out laboratory spaces**
37	3.1 Support systems
40	3.2 Services spines and bridges
42	3.3 Laboratory furniture
49	3.4 Fume cupboards and hoods
53	**4 Laboratory equipment**
53	4.1 Categories
53	4.2 Range of equipment
59	**5 Laboratory services**
59	5.1 Concepts and strategies
61	5.2 Mains supplies
61	5.3 Bench services
65	5.4 Dripcups and wastes
68	5.5 Drains
68	5.6 Power, lighting and communications
70	5.7 Safety, security and safeguarding of work and specimens
74	5.8 Environmental services
79	**6 The building fabric**
79	6.1 General
79	6.2 Partitions
81	6.3 Doors
81	6.4 Floor finishes
82	6.5 Ceilings
82	6.6 External envelope
82	6.7 Builder's work in connection
84	6.8 Builder's work ducts
84	6.9 Fire compartment penetrations
85	**7 The design team**
85	7.1 Members of the team
85	7.2 Responsibilities
86	7.3 How the team functions
86	7.4 Programmes
86	7.5 Design team meetings
87	**Technical supplements**
88	TS1 Radioactive lab classification
89	TS2 Containment of dangerous pathogens
90	TS3 Laboratory containment facilities for genetic manipulation
91	TS4 Microbiological safety cabinets
92	TS5 Clean rooms
94	TS6 Initial questionnaire for laboratory projects
95	TS7 Room data sheet
96	TS8 Laboratory furniture performance specification
100	TS9 Lab furniture menu
107	TS10 Lab furniture suppliers
108	TS11 Fume cupboard criteria
109	TS12 Fume cupboard schedule
110	TS13 Fume cupboard performance specification
113	TS14 Schedule of taps and valves
114	TS15 Bench outlets: colour code chart
115	TS16 Services into each lab module
116	TS17 A servicing concept for research laboratories
120	TS18 Chemical resistance chart
123	TS19 Radioactive shielding
124	TS20 Fire extinguishers
125	TS21 Scientific disciplines
127	TS22 External design temperatures
130	TS23 Conversion data
131	**Further reading**
132	**Index**

Acknowledgements

My thanks to Maritz Vandenberg for his encouragement and support, and for his help with the complexities of preparing matter for publishing. Special thanks to my wife Adelaine for the many hours spent over the keyboard, first in typing the original draft and then in slotting in the myriad alterations, additions and fine tuning that followed. Thanks also to Jo Stocks for typing the tables and making sure that each was contained within its allotted space.

I should like to thank University College London and another client (who wishes to remain anonymous) for permission to photograph their labs and reproduce the photos in the book.

I should also like to thank the following suppliers for their help in providing illustrations of their products, which are mentioned in the text:

- Lab-flex
- Broen Valves Ltd
- Vulcathene
- MK Electric Ltd

I am grateful to the British Standards Institute for granting permission to reproduce the information in Technical supplements 2, 3 and 5 and to *The Architects' Journal* for the text reproduced in Technical supplement 21.

Note to readers I cannot guarantee that any parts and details shown in this book will be appropriate for a particular use, and advise readers to obtain expert advice if using items not identical to those shown. In projects where I refer to particular parts and details these may differ from their original design.

A typical laboratory

Introduction

1 Introduction

1 Aims

This is not a glossy coffee table book, with dramatic views of glamorous laboratory buildings such as Louis Kahn's Salk Institute at La Jolla or Michael Hopkins' Schlumberger lab in Cambridge. Nor does it concern itself with the external appearance of laboratory buildings, which in the author's view ought like all buildings to reflect what is going on inside them: a building should say what it is.

Neither is it a book that explains esoteric scientific formulae, deciphers obscure scientific acronyms or covers laboratory practice and discipline.

This is a 'nuts and bolts' book for designers, which deals with the spatial requirements of laboratories, the means used to fit them out, the equipment that goes into them and the engineering services that they need in order to function. It is intended as a practical introduction to the design of laboratories and is therefore directed specifically towards the information needs of the design team – in particular those of the architect and building services engineer – together with those of the people responsible for briefing the team.

This book assumes that the members of the design team who use it are already experienced in the design of buildings. It assumes that the architects will be familiar with general building design, that the structural engineers will be familiar with the design of building structures, and that the building services engineers will be familiar with the design of engineering services for buildings. So the book does not concern itself with matters that are common to all other building types, nor with laboratory or scientific matters that are superfluous to the design team's needs. Its primary aim is to inform on the specific and special requirements of laboratory buildings for those coming new to the subject.

It has been structured in the sequence in which the designer would normally proceed.

- Chapter 1 includes a general introduction to the subject, relevant regulations, briefing procedures and the initial planning decisions to be taken.
- Chapter 2 examines the spaces that are found in laboratory buildings.
- Chapter 3 looks at the systems used to fit out such spaces.
- Chapter 4 covers the equipment likely to be found in them.
- Chapter 5 explains the engineering services needed in them.
- Chapter 6 lists aspects of the building fabric in which laboratories have specific performance requirements.
- Chapter 7 deals with the composition and responsibilities of the design team.
- Technical supplements 1–23 gather together information referred to in the text, which would have interrupted the flow if incorporated into the text.

The book is concerned with bench-scale laboratories, i.e. those in which the scale of activity is that associated with benches and with equipment other than benches but of a similar scale, such as refrigerators, centrifuges and fume cupboards. Most laboratories fall within this category.

The book covers the performance requirements that are of concern to those responsible for the design and building of laboratories. It aims to supply the information necessary for the newcomer to understand and interpret the data presented by a client and, where unable to do so, to ask the questions that will extract satisfactory explanations of, or amplifications to, the initial brief.

It does not provide 'off-the-peg' design solutions, but instead seeks to explain the principles that will enable designers to arrive at their own solutions. Illustrations are only included to clarify the information given in the text and not to suggest firm solutions to specific briefs. The tables, schedules and performance specifications given in the text and technical supplements show examples that have been used successfully but which may need to be modified to suit individual projects.

The book also aims to be of use to those responsible for briefing the design team, both as a guide to the nature of the information that the design team requires and as a check list.

Terms such as 'generally' and 'usually' are frequently used. This has been done deliberately in order to avoid giving the impression that the information presented is always universally applicable and immutable: there is not always consensus among scientists on all the matters covered in this book. The information given provides a sound basis for the designer to propose solutions that should usually prove acceptable, but there will inevitably be occasions when an individual scientist's practices, preferences, or prejudices will lead to the need for an alternative approach. The information is based upon experience in the design and construction of laboratories in the UK, but is applicable internationally apart from specific references to UK regulations, standards, firms and contractual practices.

When used in the text, 'laboratory' has often been shortened to 'lab', because this is the term in general use among those who work in them.

Where the term 'building contract' is used it includes the complete project: the building proper plus the engineering services.

The book has been written primarily with research labs in mind, as their requirements are less well documented than the other types, and its emphasis on the need to design for change applies particularly to research labs.

The photographs of laboratories in the book are of both new-build and refurbishment projects for which the author was responsible, and show the labs in use. The 'environment of clutter' that John Weeks (section 1.2.4) described as a characteristic of the appearance of a working laboratory is seen to be very much in evidence.

1.2 The laboratory as a building type

1.2.1 Generally

Laboratories are spaces in which tasks are performed that:

- require the presence of a variety of engineering services at each workstation in positions in which they are convenient to the user;
- may require specific environmental conditions;
- may require protective measures.

It is the combination of these three requirements that distinguishes labs from most other building types.

Laboratories may vary in sophistication from naturally ventilated, lightly serviced spaces to fully air-conditioned spaces with a heavy concentration of services. They may vary in size from a single room housing the benching, services and equipment to a multi-storey complex containing large numbers of labs together with shared ancillary accommodation.

Whatever the degree of size or sophistication, the information given remains valid and, while it is particularly relevant to research and scientific labs, in principle it applies to virtually any bench-scale lab other than a production lab, i.e. one in which commercial production processes are carried out under laboratory conditions.

The three main types of bench-scale labs are: teaching, routine and research.

1.2.2 Teaching laboratories

Teaching labs are those in which procedures and techniques are taught by demonstration and practice. They traditionally consist of a large lab area with the teacher's demonstration bench and rows of island benches for students, plus a preparation room and washing-up room. They are relatively lightly serviced and seldom change, so that flexibility is not important. The benching and equipment is similar to that used in the other two types, as are the two ancillary rooms to the others' equivalent room types.

1.2.3 Routine laboratories

Routine labs, such as hospital pathology labs, are those in which the same activities occur in the same places for long periods. They do not usually require as wide a spread of ancillary rooms nor as comprehensive a supply of special services as research labs, but increasingly as new techniques are employed and more automated equipment is used they do require more flexibility in their lab furniture than was previously the case.

1.2.4 Research laboratories

Research labs, by their very nature, are those in which it is often impossible precisely to predict the activities that will take place and in which these activities frequently change. The British architect John Weeks, after many years of laboratory work, has said of research labs:

'The most enduring requirement for an architect in the design of research facilities is the maintenance of the ability of the users to use the facility in an ill-defined way. All programmes for research facilities are out of date when the facility is brought into use. All work patterns will have changed to some degree or another. All briefs are wrong, to a greater or lesser extent. All laboratories have to be furnished in such a way that they may be stripped out, altered or marginally changed at the will of the users. All equipment used in laboratories is obsolescent and will soon be changed. Nothing should be built into the structure, because the structure has a different life span to all the functions which occur within it. The working laboratory is an environment of clutter; that is the characteristic appearance of laboratory work and an architect interferes, with his obsession for visually neat order, at his peril and the peril of his clients.'

Flexibility is therefore the most important aspect of research lab design, and the research worker should have the maximum personal control over furniture, services and environment, with the minimum need to rely on the assistance of others when these require to be rearranged or changed.

1.2.5 Ancillary rooms

In addition to the laboratories proper, other accommodation is required to provide shared specialized equipment, instruments or techniques, constant temperatures, specific environmental or protective requirements, together with washing-up and storage facilities, and staff rooms.

Lab types

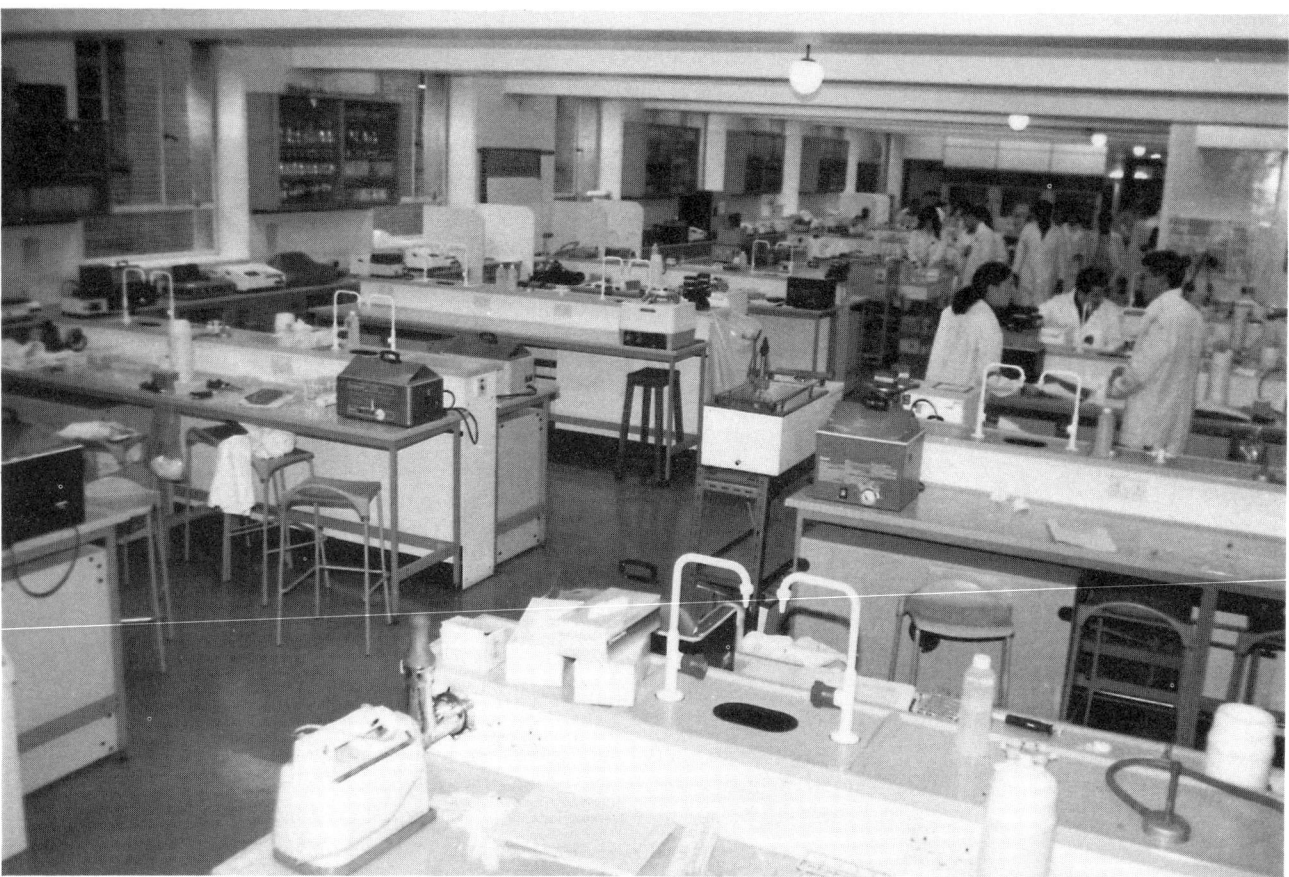

Teaching lab, with student benches

Research lab, with rig and table

1.2.6 Offices

There is considerable paperwork involved in research activities, which cannot be carried out in the labs without interfering with bench work. Offices separate from the lab are therefore required for senior staff, as close to the lab as is practicable.

1.2.7 Animal rooms

Research procedures often involve the use of animals, which require separate accommodation adjoining the labs. In the UK this is subject to very stringent Home Office regulations. It must be separate and restricted to use by designated personnel, who must be licensed by the Home Office for this work. The relevant authorities should be consulted for projects outside the UK.

1.2.8 Scientific disciplines

The three major divisions in science – chemistry, physics and biology – do not have any real relevance to the design team (any more than the old descriptions of 'wet' and 'dry' labs do), but information on these and their numerous subdivisions is given in Technical supplement 21.

1.3 Regulations and standards

Laboratories, like other building types, are subject to the usual town planning and Building Regulations requirements. In the UK, under the latter they are customarily included in the 'Offices' purpose group classification, and should be discussed with Building Control officers and the Fire Brigade as soon as a preliminary sketch layout has been prepared, to clarify means of escape and fire resistance requirements before developing the design.

In the UK the local health and safety officer should also be consulted and will be concerned primarily with the emission of any dangerous fumes and wastes and with the noise levels of plant. The client should supply the expected concentrations of fumes and fluids to be discharged into wastes, and the building services engineer should advise on plant noise. The levels of toxic discharges into the drains will also concern the responsible authority, which may be either the local water supply utility or the local authority.

A list of the major relevant statutory requirements and standards that apply in the UK is given below with, alongside, some relevant EC Directives, and reference is made in the text to such regulations when matters are discussed that are subject to their requirements. But the design team should remember that the people most familiar with the regulations that affect their disciplines are likely to be those who will work in the building that is being designed, and they should be consulted on these matters, as on all others. Above all, the design team should avoid taking decisions on scientific matters as these are beyond their competence and outside their sphere of responsibility.

UK regulations and standards and EC Directives

Statutory Regulations

The Town and Country Planning Act 1990
The Building Regulations
Health and Safety at Work etc. Act 1974
Offices, Shops and Railway Premises Act 1963
Highly Flammable Liquids and Liquid Petroleum Gases Regulations 1972
The Notification of New Substances (Amendment) Regulations 1991
The Good Laboratory Practice Monitoring Regulations 1992 (87/18/EEC; 88/320/EEC)

Radioactive standards

Radioactive Substances Act 1993
Ionizing Radiations Regulations 1985 (80/836 Euratom; 84/467 Euratom)
Design of Laboratories for Work with Radioactive Materials – Symposium by the Association of University Radiation Protection Officers 1970

Biohazardous standards

Categorization of Pathogens According to Hazard and Categories of Containment – Advisory Committee on Dangerous Pathogens 2nd edn 1990 (93/88/EEC; 90/679/EEC)

Laboratory Containment Facilities for Genetic Manipulation – Advisory Committee on Genetic Manipulation, Note 8, 1988 (90/219/EEC)
Genetic Manipulation Regulations 1989 (90/219/EEC)
The Genetically Modified Organisms (Contained Use) Regulations 1992

Animals regulations

Animals (Scientific Procedures) Act 1986
Code of Practice for the Housing and Care of Animals used in Scientific Procedures 1989 (86/609/EEC)

British Standards

BS3202 : 1991 Laboratory Furniture and Fittings
BS5295 : 1989 Environmental Cleanliness in Enclosed Spaces
BS5588 : 1983 Fire Precautions in the Design and Construction of Office Buildings
BS5726 : 1992 Microbiological Safety Cabinets
BS7258 : 1990 Laboratory Fume Cupboards

1.4 The brief

1.4.1 General

Laboratory requirements are inevitably complex, variable and confusing, and it is essential that the brief should be comprehensive and precise if the resulting design is to be successful. A bad brief is more likely to result in an unsatisfactory building than is the case with most building types.

To ensure that briefing meetings with the client are productive they need to be structured to produce the information required by the designers without time being wasted in general discussion. Some clients have their own questionnaires and pro formas for this, but questions that emanate from the designers are more likely to be presented in a form in which the answers will be of most use to the designers. The architect should therefore take the initiative in structuring the client meetings by presenting his/her own questionnaires and pro formas, examples of which are included in the technical supplements. Copies of the completed forms should always be sent to the client to confirm the information obtained.

At the first meeting with the client a questionnaire should be tabled to elicit a general statement of the requirements that will set out the overall parameters for the design. An example of such a document is shown in Technical supplement 6. At this meeting the client should also be informed of the stages in which it is proposed to obtain the brief, as suggested below.

The brief can usefully be obtained in two stages, which correspond to the stages in which the design team requires information to progress the scheme. The first is that needed to produce the initial sketch designs; the second is that needed to produce the detailed designs, from which the production information can be prepared.

1.4.2 First-stage brief: accommodation requirements

This should include:

1 *accommodation schedule* – a room-by-room list giving name, area and occupancy of each;
2 *room relationship statement* – guidance on rooms that need to be grouped, in close proximity or *en suite*;
3 *operational policy statement* – a general explanation of how it is proposed to operate the building, e.g. where staff will enter, how the building will be supplied, how waste will be be disposed of, hours of operation;
4 *general environmental conditions* – whether mechanical ventilation or air conditioning is required in specific areas or throughout;
5 *non-standard requirements* – identification of any rooms or areas in which out-of-the-ordinary space, servicing or other demands will affect the building form.

1.4.3 Design concept

Before any design work is commenced on the building proper a written statement should be prepared that establishes the principles upon which the design of the building will be evolved – the design concept – and agreed with the client. In a laboratory building this should include the lab module on which planning is based, the plan form to be adopted, the structural system, the servicing strategy, where the bench outlets will occur and how they will be supplied, the policy for removing fume cupboard fumes and on supplying make-up air to rooms with fume cupboards, the position of the plant room, the type of laboratory furniture to be used, etc. This will be a design team document (usually prepared by the architect), in which all members will be ready to support the proposals for their particular speciality.

1.4.4 Second-stage brief: detailed requirements

This should include room-by-room details of:

1. *engineering services requirements* – e.g. power supplies, water supplies, special gases;
2. *environmental requirements* – ventilation, temperature, humidity, lighting;
3. *fittings and equipment* – e.g. benching, cupboards, fume cupboards, equipment;
4. *finishings* – floor, wall and ceiling finishes.

1.4.5 Room data sheets

Information on the second stage is best gathered by means of room data sheets, on which the detailed requirements and content of each room are recorded on a standardized pro forma for the project. An example is included in Technical supplement 7. Many client bodies will have their own versions of these. Room data sheets are very useful, but they require interpretation and should not be taken at face value. To be applicable to a wide range of projects they usually offer a comprehensive range of choices and, when faced with such a 'shopping list', there is sometimes an understandable tendency on the part of those completing them to indicate items that may not be essential, may be difficult or expensive to provide, or may even be mutually exclusive! The best means of interpreting the information on the sheets is to discuss them direct with the end-users: the people who will actually work the labs are not always the ones who completed the sheets.

This discussion is not always possible. On new projects starting from scratch it is unlikely that any staff will have been appointed other than perhaps the head of the new unit and possibly his or her deputy, so that the detailed requirements of the staff who will work the labs are not yet available: another reason for the need for maximum flexibility in the design.

Existing laboratory units normally have a major-domo figure, usually a senior lab technician, with responsibility for ensuring the proper working of the facilities in each lab, and to whom the lab users have recourse whenever they require modifications to their labs or when any of their lab facilities is not functioning as intended. The senior technician is likely to have a much better practical knowledge of the items that affect the design team than the users themselves and so should be included in the briefing group at the second stage.

1.4.6 Floor data plan

Receipt of the second-stage information makes it possible to produce the 1:50 scale floor data plan(s), which show all services, spines, benches, bench services, fume cupboards and all floor-standing (i.e. space-consuming) equipment and furniture, with the equipment's servicing requirements.

When the floor data plan is approved by the client it will show the definitive brief, which will provide the building services engineer with the information required to commence the detail design work and the information required

Floor data plan.

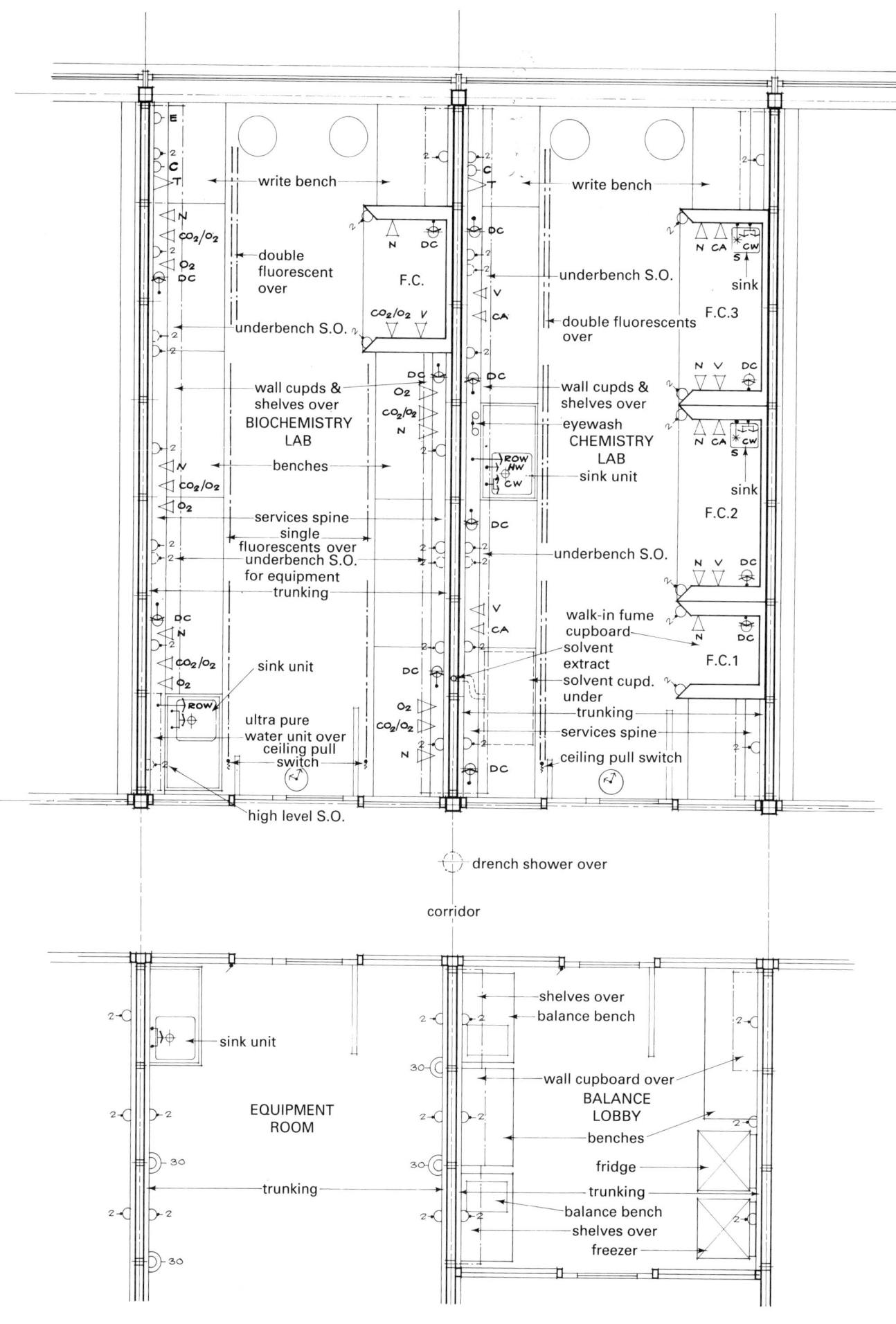

Introduction

to prepare an accurate estimate of cost (the responsibility of the quantity surveyor in the UK).

Revisions and modifications to the brief will continue to occur as the scheme is developed, and these should be shown on the floor data plan, which should be updated to reflect the current state of the brief, against which production information can be checked.

Although it should never be issued for construction purposes the floor data plan can usefully be issued to the contractor on an 'information only' basis, as it is usually the only plan that shows on one drawing all the fittings, bench services, etc. that occur in each room.

1.5 Initial procedures in planning laboratories

1.5.1 Modular planning

The vast majority of activities that take place in laboratories can perfectly well be accommodated within a common room-size and room-shape – the 'lab module' – or in multiples of that module, with the differing requirements of individual labs met by modifications or additions to the fitting-out or servicing of the basic lab module (just as car manufacturers provide a basic model on which they offer optional extras such as choice of engines, power steering, or sun roof).

The aim therefore should be to minimize the need for non-standard, closely tailored accommodation, and to rationalize room area requirements into a lab module suitable for the project and then to obtain the client's agreement to that module (for a basis on which to choose a suitable module refer to Chapter 2). Once this has been done, then using the total laboratory area requirements from the accommodation schedule the number of lab modules needed can be established.

1.5.2 Plan form

As a general rule the lab module accommodation will be continuously occupied and so should be positioned in naturally lit, 'prime space' areas. Individual labs may consist of one or more lab modules. To provide the flexibility to vary the number of modules in a lab as needs change it is therefore desirable to arrange lab modules side by side without other intervening accommodation that could inhibit this flexibility.

Non-standard labs generally require mechanical ventilation to meet their performance requirements and are not usually continuously occupied. They therefore need not be positioned in prime-space areas and may be placed in internal areas.

Ancillary rooms seldom need to change in form, their functions often either require the exclusion of natural light or do not need it, their construction may militate against change, and they are not usually continuously occupied. It is therefore logical to separate them from the prime-space lab areas and relegate them to internal areas, where they will not inhibit the flexibility of the lab modules.

There are four types of basic plan form suitable for a laboratory building and most lab buildings are variations of these. They are as follows.

Type 1

Single banks of accommodation on both sides of a single corridor. Both banks will be naturally lit and both may be lab module depth to accommodate labs, or one may be shallower to house special labs and/or ancillary accommodation.

Type 2

Accommodation double-banked on one side of a single corridor and single-banked on the other. The single bank on one side and the outside bank on the other will usually be of lab module depth and naturally lit, with the inner bank shallower and mechanically ventilated, to house special labs and/or ancillary accommodation.

Type 3 Accommodation double-banked on both sides of a single corridor. Both outer banks are of lab module depth and naturally lit; the inner banks may be shallower and will be mechanically ventilated, to house special labs and/or ancillary accommodation.

Type 4 Two corridors, with naturally lit lab accommodation single-banked on the outside of each and the core space between the corridors mechanically ventilated to house special labs and/or ancillary accommodation.

The decision on which type to adopt for a project will depend upon its briefing requirements and the size of the site that is available. Type 3 occupies most depth, followed by Type 4, Type 2 and Type 1, in that order. Types 2, 3 and 4 have internal rooms that require mechanical ventilation. As a general rule Types 2, 3 and 4 are more suitable than Type 1 for labs with a good range of ancillary accommodation.

The decision on the plan form to be used should be made at design concept stage.

1.5.3 Vertical ducts

For ease of initial installation together with subsequent addition and modification, it is desirable to have vertical ducts at the end of each side of the lab module to house submains that feed direct into the bench services: the piped services run-outs and the power trunking.

Though they may be quite small in area these ducts are relatively closely spaced (3.3–3.6 m (11–12 ft) centres) and are therefore space-consuming.

Their position will depend upon which of the four options shown in Chapter 5, section 5.3.1 is chosen.

If the ducts are to accommodate fume cupboard extracts then their size will increase substantially, for in place of the relatively small-bore piping required for the submains they will be housing extract ducts of at least 300 mm (12 in) diameter.

The author has used option 3 (ducts at both ends) successfully, using shallow ducts at the corridor end for submains (where frequent access is required) and deeper ducts at the window end for fume cupboard extracts (which seldom require access and where the extra depth occurs outside the main envelope of the building).

An alternative strategy for accommodating submains together with extract and supply ducts outside the user spaces of the building is given in Technical supplement 17.

Information on vertical ducts is given in Chapter 6, section 6.8.

The decision on which vertical duct position is to be adopted should form part of the design concept.

In single-storey projects the vertical ducts are not normally required because there is no user accommodation above the labs and the submain drops can therefore emerge direct into the lab from the plant room or roof space over, to drop down within the support/partition zone (Chapter 3, section 3.1); fume cupboard extracts can similarly emerge from the top of the fume cupboard direct up into the plant room or roof space.

Plan forms.

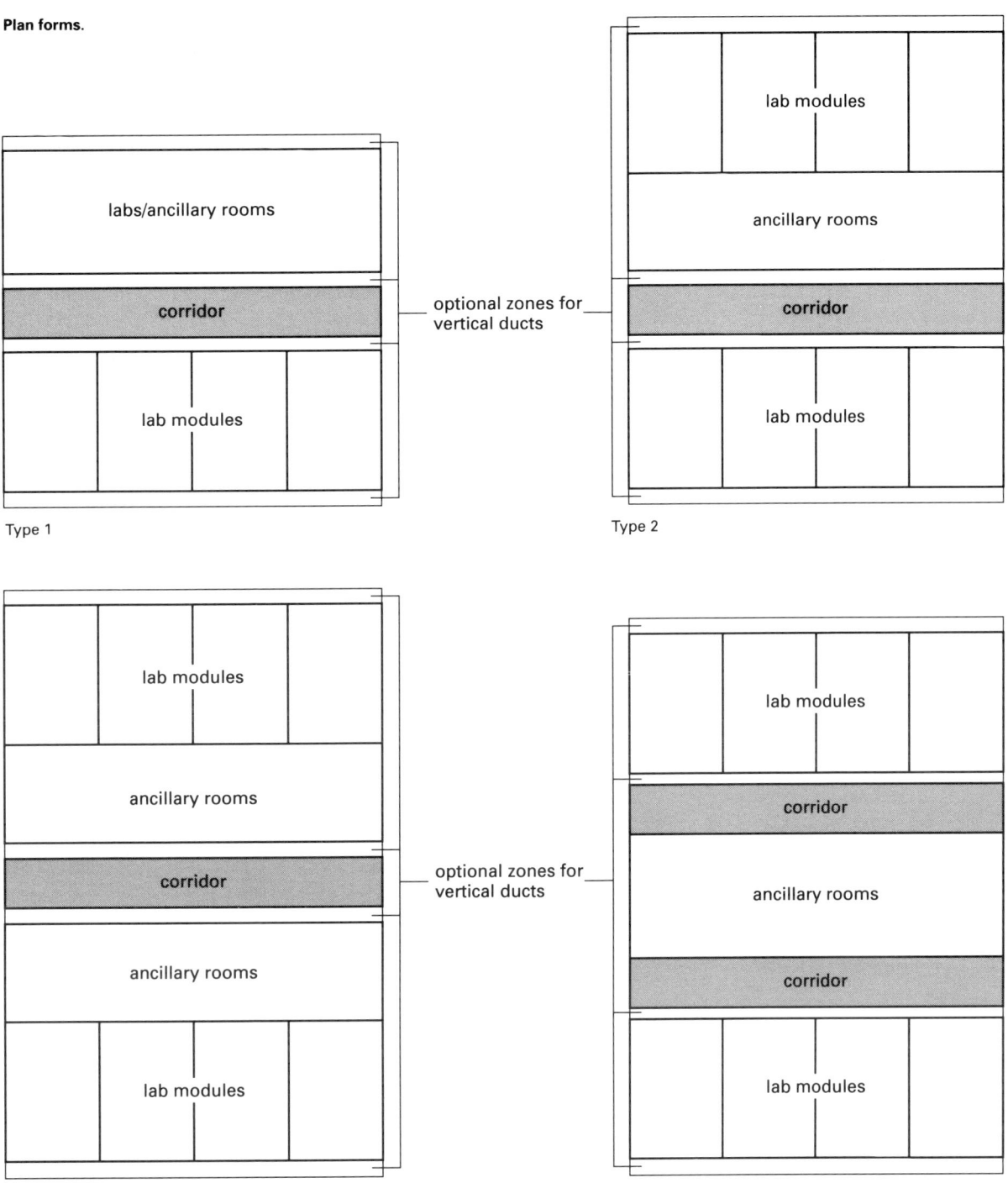

Type 1

Type 2

Type 3

Type 4

Initial procedures in planning laboratories

1.5.4 General planning

Within the chosen plan form the planning of laboratory buildings does not diverge from the planning principles that apply to all building types, and the accommodation is usually disposed as follows:

- main entrance giving access to an entrance hall/foyer with reception desk and waiting area for visitors;
- administration areas (e.g. reception/secretaries' offices, director's office), toilets, common room and seminar room, all convenient to the entrance hall (and including stair and lift where applicable);
- lab corridor(s) accessible from the entrance hall, with lab modules and ancillary accommodation arranged about the corridor(s) in the configuration of the chosen plan-form type;
- service entrance (for deliveries, waste disposal and access to external stores) into a service lobby, the latter with direct connection to a lab corridor;
- animal house entrance (where applicable) direct off a lab corridor and convenient to the labs that use the animals, into a lobby off which staff changing rooms and toilets, offices and rest room are accessible. First section of animal house corridor, accessible from the lobby, containing bedding and food stores, cage wash and any other ancillary rooms. Second section of animal house corridor, leading on from first, with animal rooms on both sides.

2 Laboratory spaces

2.1 The laboratory space

2.1.1 Standard laboratories

By 'standard' is meant those labs that can be accommodated within the standard module, in which the differences in performance requirements between individual labs can be met by variations to the servicing, lab furniture or finishes within the lab module, and which in planning the laboratory building can therefore all be included in the 'lab module depth' accommodation in the basic plan form types described in Chapter 1, section 1.5.2.

The three major divisions in science – chemistry, physics and biology, together with their numerous subdivisions – all fall within this category, as do virtually all other bench-scale research laboratories apart from those in the 'specialist' category below. The differences in performance requirements should be spelled out in the briefing information for individual rooms.

The laboratories in this category are continuously occupied throughout the working day (and in research labs often at various times beyond the working day) and are normally planned in naturally lit prime-space perimeter areas.

2.1.2 Specialist or non-standard laboratories

General

By 'specialist or non-standard' is meant labs whose performance requirements are such that they may not be able to be accommodated within the standard lab module. There are three main subdivisions:

- labs in which the levels of radioactivity released require protective measures to be taken;
- labs in which the presence of dangerous pathogens requires protective measures to be taken;
- labs in which the cultivation of tissue cultures requires a controlled environment together with protection for the cultures against the ingress of contamination from outside the labs.

All three essentially provide specialist facilities for standard labs; they are generally not continuously occupied and so may be planned in internal core areas.

Radioactive laboratories

In the UK the statutory requirements for these are set out in the Radioactive Substances Act 1993 and the Ionizing Radiations Regulations 1985, but these do not offer practical guidance to lab designers. The British Radiological Protection Association held a Symposium on the Design of Laboratories for Work with Radioactive Materials in 1970, which provided such guidance. It graded labs in ascending order of sophistication as follows (refer also to Technical supplement 1 for further information).

Standard labs

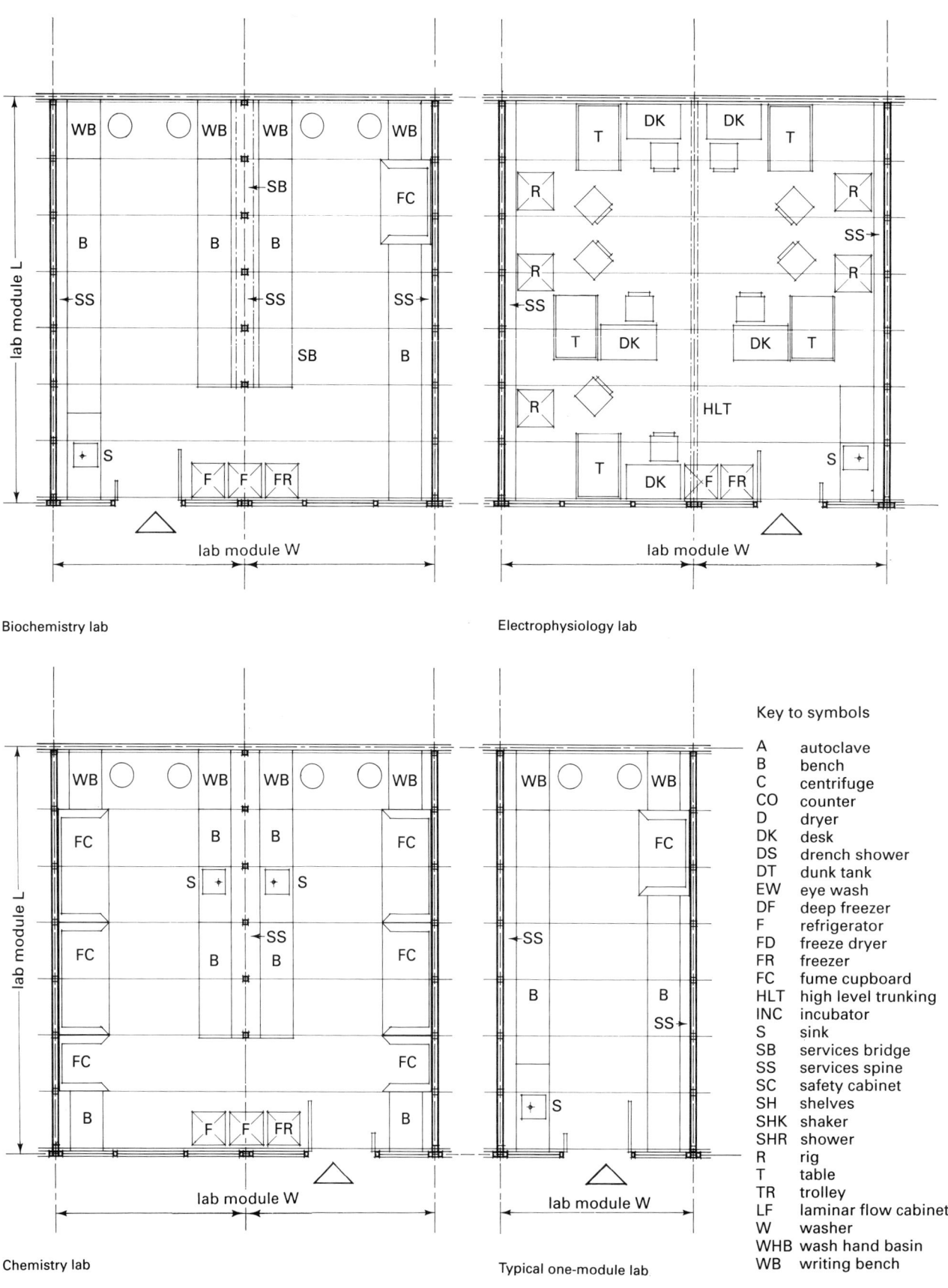

Biochemistry lab

Electrophysiology lab

Chemistry lab

Typical one-module lab

Key to symbols

A autoclave
B bench
C centrifuge
CO counter
D dryer
DK desk
DS drench shower
DT dunk tank
EW eye wash
DF deep freezer
F refrigerator
FD freeze dryer
FR freezer
FC fume cupboard
HLT high level trunking
INC incubator
S sink
SB services bridge
SS services spine
SC safety cabinet
SH shelves
SHK shaker
SHR shower
R rig
T table
TR trolley
LF laminar flow cabinet
W washer
WHB wash hand basin
WB writing bench

Laboratory spaces

Specialist labs

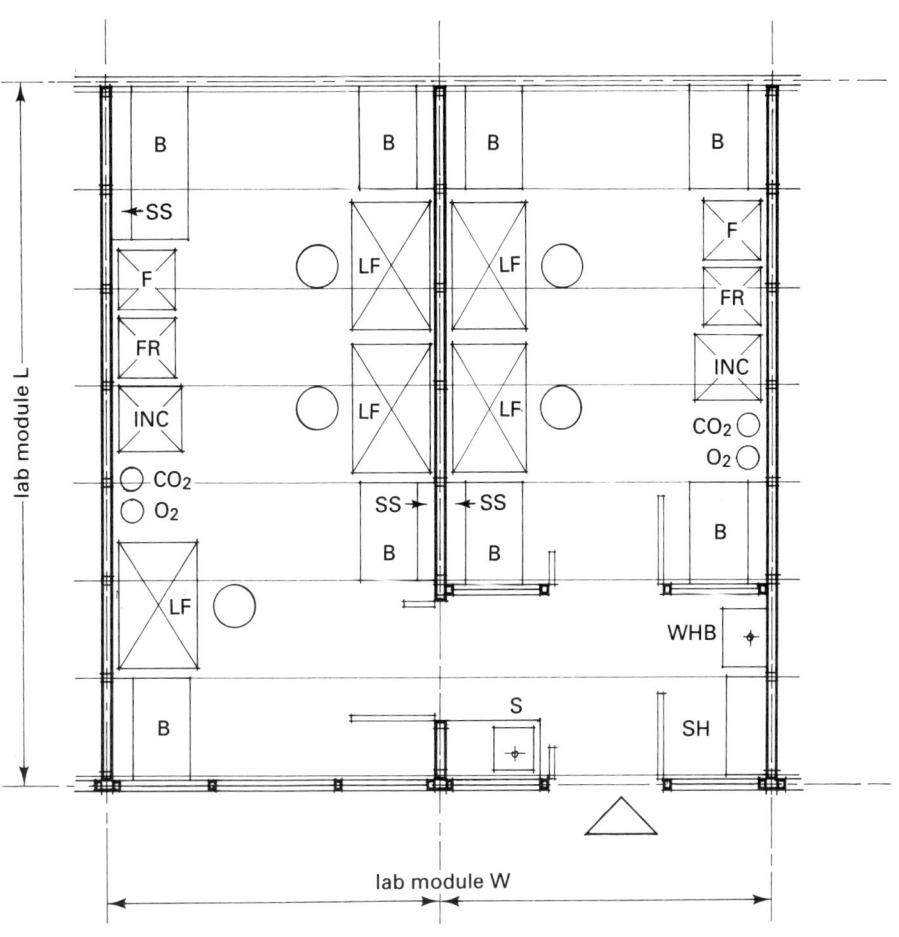

Tissue culture suite

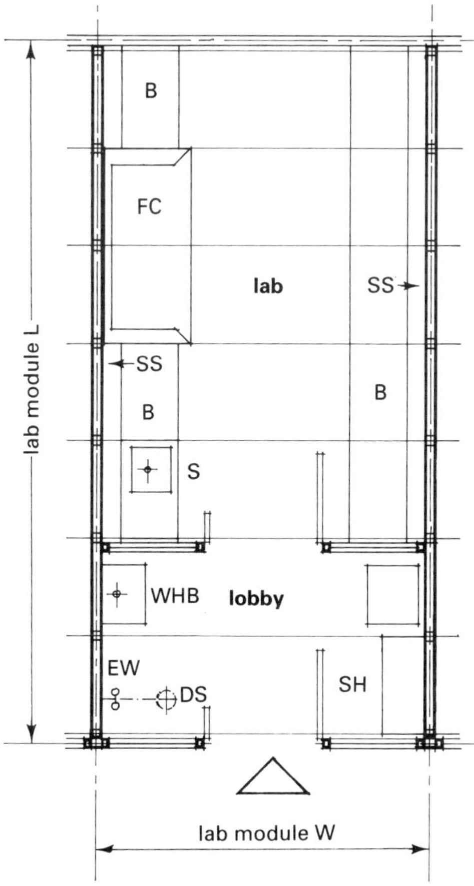

Radioactive lab

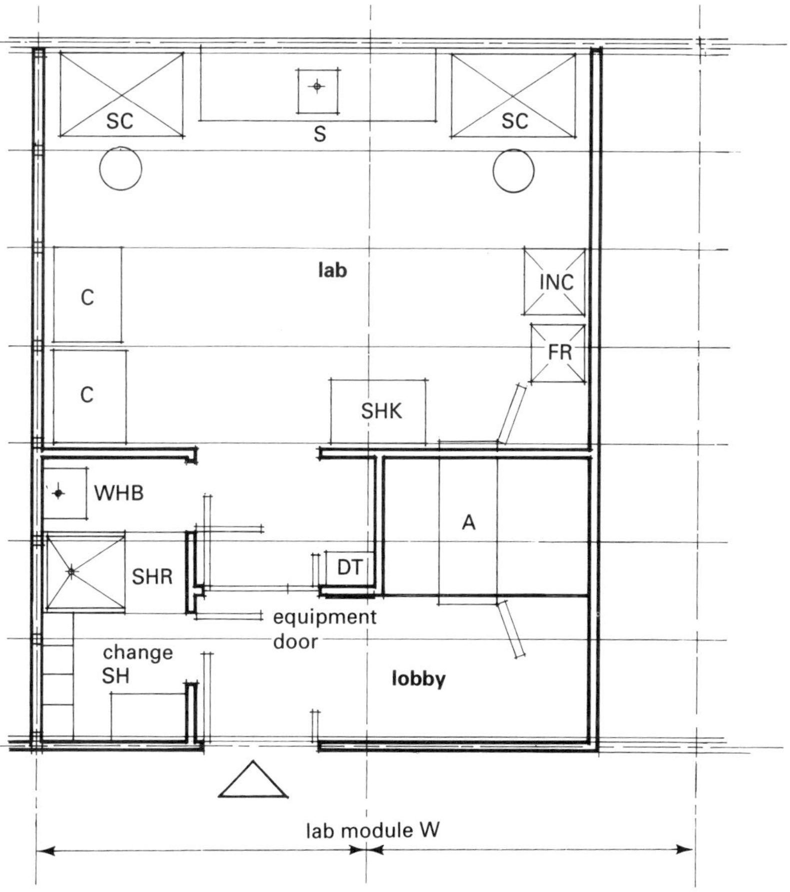

Biohazardous lab

The laboratory space

- *Grade C* – Most new laboratories will be suitable for this category, e.g. welded sheet PVC flooring and laminate-covered worktops, and this grade therefore falls within the 'standard' lab category.
- *Grade B* – This is the usual grade envisaged when labs are described as 'radioactive', and a cut-off lobby is usually provided. A fume cupboard must be provided and a wash handbasin with elbow-operated taps is required. Walls, floors, ceiling, lab furniture and fittings must be smooth and impervious, with gaps sealed. Welded sheet PVC flooring and epoxy resin benchtops with raised edges are usual.
- *Grade A* – Access is through a fully equipped change room and barrier, most work is done in glove boxes, the highest standard of finishes is required, with all joints carefully sealed, and a plenum mechanical ventilation system is fitted with exhaust system filtered. The walls, floor and ceiling must be sufficiently dense to shield the occupants of adjoining rooms from the radioactive material held in the room. Epoxy resin flooring and coved skirting with sprayed plastic finish to walls and ceiling are usual.

The brief should specify the grade of any radioactive lab, the scale within that grade and the precise requirements. For information on radioactive shielding refer to Technical supplement 19.

Biohazardous (dangerous pathogens) labs

In the UK the report of the Advisory Committee on Dangerous Pathogens 1990 categorized dangerous pathogens into hazard groups 1–4 and categories of containment into levels 1–4. The latter correspond to the hazard groups and set out the requirements that concern the building designer. These are given in Technical supplement 2, but are summarized as follows.

- *Level 1* – No requirement that could not be met within the standard lab classification.
- *Level 2* – 24 m^3 (850 ft^3) per worker; wash handbasin with elbow-action taps; an autoclave for sterilizing waste materials readily accessible in the same building. A Class 1 microbiological safety cabinet may be required. These requirements should usually be able to be met within the standard lab classification.
- *Level 3* – 24 m^3 (850 ft^3) per worker; sited in an area away from general circulation (i.e. with cut-off lobby); stringent airflow requirements; wash handbasin with elbow-operated taps; an autoclave either within the lab or in the lab suite; all lab procedures with infected materials to be conducted in a Class 1 or Class 3 microbiological safety cabinet with stringent exhaust requirements; the lab to contain its own equipment (centrifuge, incubator, etc.). These requirements are unlikely to be met within the standard lab classification.
- *Level 4* – The lab unit must be a separate building or form an isolated part of a building. Entry through an airlock with changing and showering facilities; independent mechanical ventilation system to maintain an airflow from the outside area into the lab (negative pressure); double-sided autoclave from lab to clean lobby; double-ended dunk tank from lab to clean lobby; all lab procedures conducted in a Class 3 microbiological safety cabinet.

Laboratory containment facilities for genetic manipulation

In the UK the Advisory Committee on Genetic Manipulation has produced a report, Note 8 of June 1988, which lists four containment levels with a table giving a summary of the requirements, reprinted in Technical supplement 3.

Tissue culture labs

Tissue culture facilities vary considerably and will depend upon the particular client requirements. They may range from a single room with laminar-flow cabinets and incubator, refrigerator and freezer, approached via an isolating lobby, to a suite of rooms with separate preparation room and hot room in addition to the lab(s) proper, all approached via an isolating lobby. The cultures are grown in carefully controlled conditions and are vulnerable to contamination, and the rooms are usually maintained under positive

UK SPACE STANDARDS

1 Schools

Level of work	Area per workplace (m²)	(ft²)	Scale of work	Area per workplace (m²)	(ft²)
General science	2.8	30	Bench scale	3.2*	34*
Individual projects	3.6	40	Workshop scale	4.6*	50*

* These allowances include storage and preparation.

2 Polytechnics and colleges of further education

Level of work	Area per workplace (m²)	(ft²)	Addition for ancillary rooms (%)	Addition for balance areas (%)
General science	4.6	50	15	40
Advanced science and engineering	5.6	60	25	40

3 Research (government and commercial)

Likely range	Area per workplace (m²)	(ft²)
Chemistry	8–12	86–130
Physics	6–8	65–86
Biology	6–8	65–86

Area requirements can vary considerably, depending on individual requirements and equipment. Additional area percentages from p. 18 may be used as a guide but must be verified for each project.

Space between benches

a One worker, no through traffic — 975 mm/ 3ft 2 in to 1200 mm/4 ft)
b One worker plus passage way — 1050 mm/ 3ft 6 in to 1350 mm/ 4 ft 6 in
c Two workers, back to back, no through traffic — 1350 mm/ 4ft 6 in to 1500 mm/5 ft
d Two workers, back to back plus passage way — 1650 mm/ 5ft 6 in to 1950 mm/6 ft 6 in
e Gangway only, no working spaces either side — 900 mm/ 3ft to 1500 mm/5 ft

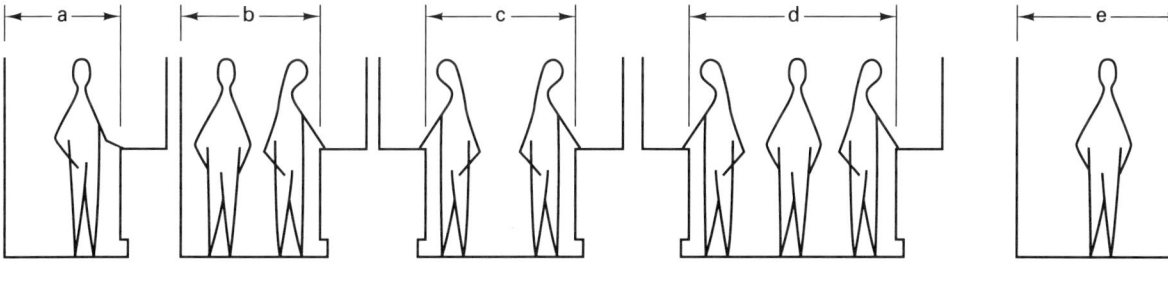

The laboratory space

4 Universities

	Category	Area per workplace		Addition for storage and prep rooms (%)	Addition for other ancillary rooms (%)	Balance area addition (%)
		(m²)	(ft²)			
Pure sciences	*Teaching labs*					
	Biological sciences (general purpose)	4.0	43	15	Ad hoc in accordance with needs (say 15%)	30*
	Biological sciences (other than general purpose)	5.0	54	15	"	30*
	Physics	5.0	54	15	"	30*
	Chemistry	5.0	54	15	"	30*
	Research labs					
	Individual or advanced	11.0	118	15	"	30*
	MSc courses	7.5	81	15	"	30*
Other technical and scientific subjects	*Teaching labs*					
	Elementary or intermediate	3.7	40	15	15	45†
	1st and 2nd honours and general	4.2–4.6	45–50	15	15	45†
	Final-year honours	5.6–6.5	60–70	15	15	45†
	Research labs					
	Research students in groups of 4 or more	7.4	80	15	15	45†
	Individual or advanced research	11.0	118	15	15	45†

* Additional balance area allowances will be needed for plant rooms, ducts, boiler houses and entrance halls: physics – up to 12½% of workplace, storage/prep and ancillary areas; chemistry and biological sciences – up to 20% of workplace, storage/prep and ancillary areas.

† Balance area % includes allowance for plant rooms, etc.

Areas for medical research labs

Space category	Area per workspace	
	(m^2)	(ft^2)
Lab areas (includes writing areas)	**11.0**	**118**
Ancillary areas (30%)	**3.4**	**37**
Total	**14.4**	**155**

- The figures for area per workspace apply to each member of the scientific and technical staff, but not to secretarial/clerical staff, animal technicians nor maintenance staff.
- An additional 18.5 m^2 (200 ft^2) is allowed for an office for the head of the unit.
- An additional area is allowed for library/seminar/coffee; a single room 18.5–28 m^2 (200–300 ft^2) in smaller units, separate rooms in larger units.
- Extra space is allowed for highly specialized facilities such as electron microscope suite, tissue culture suite, computer suite.
- Balance areas: a percentage of the total of the usable areas above should be added for circulation, toilets, plant rooms, ducts, etc. This can vary between 30% and 45%, depending upon the extent of servicing.

The laboratory space

pressure to prevent the ingress of contamination from adjoining areas. Standard finishes and lab furniture are usually acceptable, but in certain instances epoxy resin flooring and sprayed plastic wall finish may be required.

2.1.3 Space standards

The space standards that are available usually define lab areas by means of an allocated area per workspace, to which are added percentages for ancillary rooms and balance areas. Labs and ancillary rooms are classed as 'usable areas', the balance areas being made up of circulation space (corridors, stairs, lifts), toilets, plant rooms, ducts, etc.

In the UK most official organizations publish tables giving their area allowances, and some of these are shown in the attached tables. Space standards for hospital pathology labs in the UK are covered in Department of Health and Social Security Hospital Building Note No. 15.

The figures given in the tables for teaching, and research associated with teaching, provide firm information because they are derived from experience of functions that have not changed significantly.

However, the figures for independent research should be used for guidance only because the space requirements of research labs can vary considerably, depending upon the type of research undertaken. Private commercial firms usually have their own standards to suit the work that they carry out, and these will normally be included in the brief.

In the UK the Nuffield Foundation published data in 1961 suggesting that a space 3–4 m (10–13 ft) wide × 6 m (20 ft) would be satisfactory for most research labs, with a unit size of around 24 m^2 (260 ft^2) serving four, three or two workers, depending upon the work. This gives areas per workspace of 6 m^2 (65 ft^2), 8 m^2 (86 ft^2) or 12 m^2 (130 ft^2), which subsequent experience has tended to confirm. The lab module proposed in the following section reflects this experience.

A clear height in bench-scale labs of 2.7 m (9 ft) will satisfy most requirements including fume cupboards, and if a greater height is required for rigs or equipment then this should be specified during briefing.

2.1.4 The lab module

The lab module criteria given here apply to normal bench-scale laboratories. They do not apply to teaching labs (where larger areas are required) nor to projects that consist primarily of non-standard labs, such as high-risk biohazardous containment facilities; for these a module should be chosen that suits the lab unit size called for in the brief.

It is not always possible in refurbishment work to provide the lab module dimensions proposed here. Existing buildings are often too shallow to accommodate the desired depth, in which case this will have to be reduced and fewer users can be accommodated. But the minimum width shown is critical; reduction of this width will mean clear spaces between benches that would not normally be acceptable.

The traditional laboratory shape is rectangular, with benching along the long walls, the entrance door from the corridor in one short wall and a window in the other short wall. This shape gives the longest run of benching per square metre, provides an economical plan shape and the minimum of external walling, and is still the most widely used.

But research is no longer as bench-bound as previously: there is more use of plug-in electronic equipment, and work may be done by the researcher sitting in a U-shaped array of instruments. For this pattern of use a square shape is more suitable.

Lab module

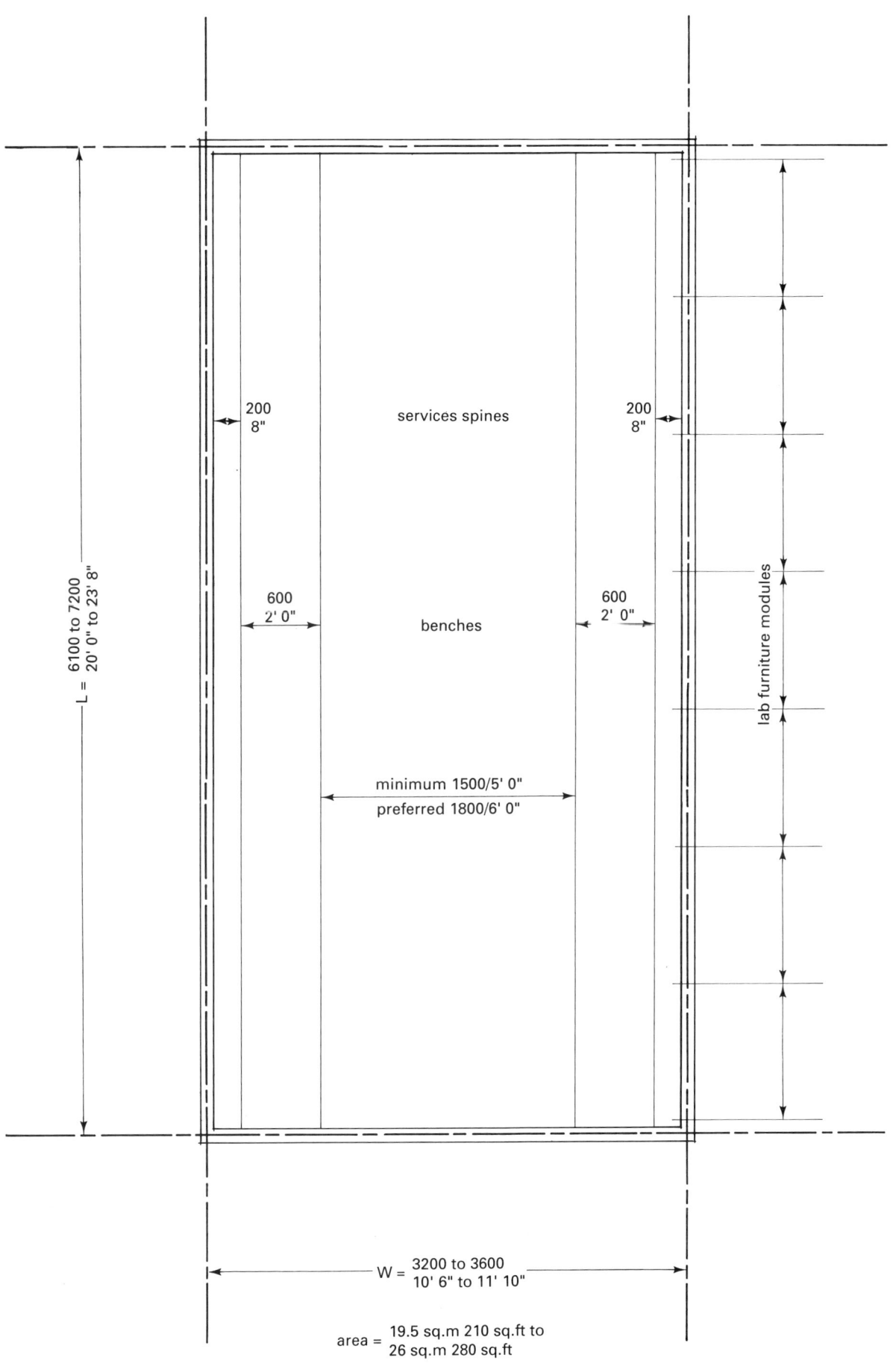

The laboratory space

The width of the rectangular shape is determined by the depth of the service spines along each wall (200 mm (8 in) each), the depth of the benches (usually 600 mm (2 ft) each), plus the clear space between (minimum 1500 mm (5 ft), preferably 1800 mm (6 ft)). These give a minimum width of 3100 mm (10 ft 2 in). Some procedures require benches 800 mm (2 ft 8 in) deep, giving a width of 3500 mm (11 ft 6 in). The addition of partition widths would add 100 mm (4 in) to these, giving a module width between 3200 mm (10 ft 6 in) and 3600 mm (11 ft 10 in). The length of the rectangular shape will depend upon the number of people to be housed, but it has been found that two parallel benches of 6000 mm (19 ft 9 in) each, including a 1500 mm (5 ft) fume cupboard and 1500 mm (5 ft) sink unit, is satisfactory under most conditions. A minimum length of 6000 mm (19 ft 9 in) should therefore be acceptable, with a maximum in the order of 7000 mm (23 ft). The addition of partition widths would add 100 mm (4 in) to these, giving a module length between 6100 mm (20 ft) and, say, 7200 mm (23 ft 8 in). These module widths and lengths give a lab module area between 19.5 m^2 (210 ft^2) and 26 m^2 (280 ft^2).

Adoption of the rectangular module enables square lab spaces to be provided by combining two modules and omitting the intervening partition.

The length of module adopted should be a multiple of the chosen lab furniture module (Chapter 3, section 3.3.1).

2.1.5 The structural bay

The width of the structural bay should relate to the width of the lab module, being one module wide or two modules wide, depending upon the structural frame system that is chosen.

The length of the structural bay for the prime-space lab areas will normally be the length of the lab module, but may also be the lab module plus the corridor width.

The length of the structural bay for the core in a core area plan may be the same as the lab module length, or that length plus corridor widths, or may be a length of its own without reference to the lab module length.

In the UK, BS6399 : Part 1 : 1984 Code of Practice for Dead and Imposed loads states that the imposed load (the load assumed to be produced by the intended occupancy of use, excluding partitions and finishes) for laboratories is to be 3.0 kN/m^2 (63 lbf/ft^2), but as a guide for bench-scale labs an overall allowance of 5.0 kN/m^2 (105 lbf/ft^2) plus point loads between 3.6 kN and 5.0 kN (810 lbf and 1125 lbf) has generally been found to be adequate, together with an allowance of 0.25 kN/m^2 (5 lbf/ft^2) for suspended services.

2.2 Ancillary spaces

This section covers the ancillary areas that are most likely to be encountered in laboratory projects. However, areas with highly specialized and unusual requirements, on which there is no previously recorded information, are sometimes called for in a brief. In such instances it is necessary to obtain the information in direct discussions with the users, who should state the activities that will take place, how these are to be carried out, the environmental conditions necessary for their successful operation, and the services that they need in order to function properly.

2.2.1 Equipment rooms

To house shared floor-standing equipment such as centrifuges, counters, deep freezers and freeze dryers. There is no standard size of room. This will depend upon the equipment to be accommodated, which should be specified in the brief. The rooms typically have a small sink unit, with power trunking around the walls. In addition to 13 A socket outlets, in the UK 20 A and 30 A outlets may be required for the centrifuges. If the room is internal then mechanical ventilation will be needed and cooling may be required. The doors should be 1½ leaves wide.

2.2.2 Instrument rooms

To house shared bench-mounted instruments such as balances and microscopes. The size of room will vary with the instruments to be accommodated and should be specified in the brief. The room typically has benches along the walls with power trunking over. If the room is internal then mechanical ventilation will be needed. Door leaves should be minimum 900 mm (3 ft) wide.

2.2.3 Cold rooms

Are usually either +4 °C (39 °F) rooms in which it is possible to work for limited periods (essentially low-temperature laboratories) or very low-temperature cold stores, usually at 220 °C (24 °F). To minimize temperature loss when the door is open it is usual for cold stores to be entered either from a +4 °C (39 °F) room or from an insulated lobby. Cold rooms will usually have insulation on the walls, ceilings and floor, that on the floor providing a threshold step against which the insulated outward-opening door closes to provide a seal. If a threshold step is not acceptable, the structural floor must be dropped over the area of the cold room, a ramp introduced or the compression seal at the threshold omitted. When +4 °C (39 °F) rooms are on the ground floor the floor insulation is often omitted.

Doors opening onto a main corridor should be recessed to avoid restricting the corridor when open; doors to +4 °C (39 °F) rooms are usually fitted with a vision panel.

Room sizes will be given in the brief and to these must be added the insulation thickness: usually about 100 mm (4 in) for +4 °C (39 °F) rooms and 175 mm (7 in) for 220 °C (24 °F) rooms. Ceiling heights of about 2200 mm (7.25 ft) are usual.

+4 °C (39 °F) rooms will have benches with an inset sink bowl with two cold-water taps, plus shelving and socket outlets on the wall above benching; 220 °C (24 °F) rooms have shelving only.

Each cold room will have a condensing unit in the open air connected with refrigerant pipes to a fan-coil cooling unit mounted at high level in the cold room.

Each room should have an emergency alarm push connected to a warning light or audible alarm, and have a temperature recorder outside the room.

The working temperatures and permissible fluctuation limits should be specified in the brief.

Cold rooms are always constructed by specialists, who fix finished insulated wall, ceiling and floor panels within the shell formed by the main contractor, will include the doors and may include shelving, lighting, socket outlets, alarms and temperature recorders. Benching to +4 °C (39 °F) rooms is best supplied by the lab furniture supplier.

2.2.4 Hot rooms/constant temperature rooms

Are essentially similar to cold rooms, but heated instead of cooled (+37 °C (98.4 °F) is common), are also constructed by specialists, and the size and temperature required, with permissible fluctuation limits, should be specified at briefing. The rooms will normally contain benching and shelving and should have alarms similar to those for cold rooms.

Constant temperature (CT) rooms are of similar construction, but their temperature variation needs may range from temperatures below the ambient to greatly in excess of ambient. They may therefore require both cooling (like cold rooms) and heating (like hot rooms).

2.2.5 Darkrooms

For processing exposed film. A size should be given in the brief, which should also state whether the room must be able to be entered while processing is in progress. If this is a requirement then either a light-tight lobby or a proprietary light-tight revolving door must be provided. If not required then a standard internal door with compression seals to all edges is usually satisfactory. Transfer grilles in the door must be darkroom type. A suitable

Ancillary rooms

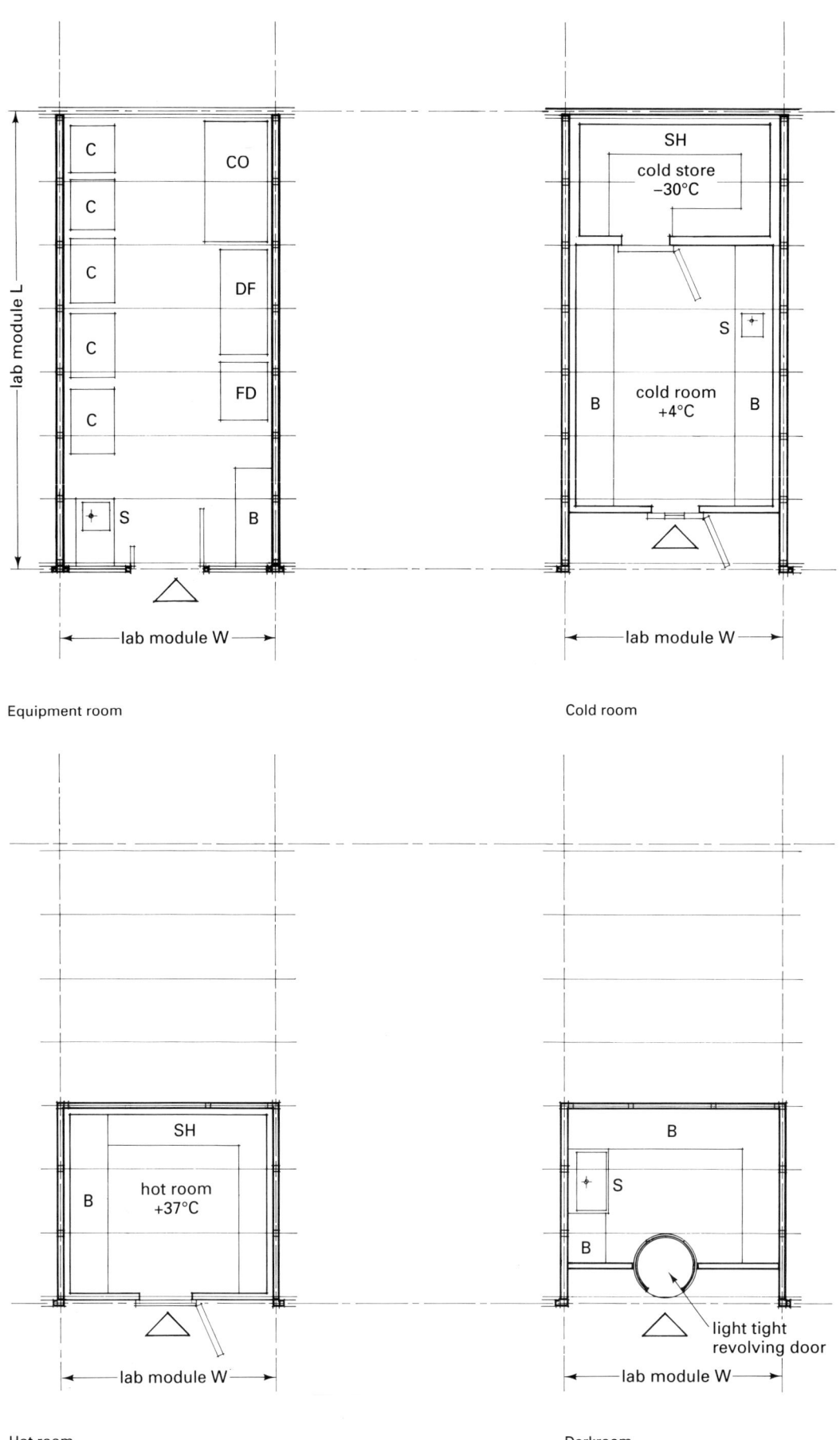

Equipment room

Cold room

Hot room

Darkroom

revolving door is supplied by Marrutt. A large fireclay sink bowl with cold-water tap and hot and cold mixer tap is required, together with wall benching. Wall-mounted socket outlets and safelights, ceiling-mounted safelight(s) and standard lighting are required. A red warning light should be provided outside the room to indicate that darkroom procedures are in progress.

2.2.6 Wash-up rooms

For centralized washing, drying and sterilizing of glassware. The size of room will depend upon the equipment required and should be given in the brief. The rooms usually contain a washing machine, dryer, autoclave and sink. In most cases one of each will be sufficient, and these can be accommodated in an area of about 20 m² (215 ft²), which allows space in front for loading/unloading the equipment and for trolley parking. Good extract ventilation is required and the door should be 1½ leaves wide. A floor gulley is required adjacent to the autoclave.

2.2.7 Electron microscope rooms

These are normally part of a suite comprising the microscope room, a preparation room and a small darkroom. The size will depend upon the instrument to be installed and should be given in the brief, but as a guide the following areas have proved satisfactory: microscope room 14 m² (150 ft²), prep room 19 m² (205 ft²), darkroom 7 m² (75 ft²). The microscope room must be able to be blacked out and to have cooling. Both the microscope and prep rooms should have doors 1½ leaves wide.

2.2.8 MRI room

Rooms housing magnetic resonance imagers, which are computers linked to superconductive cylindrical magnets into whose magnetic fields the objects to be examined are inserted. The area of the room will depend upon the equipment to be used. A room of 20 m² (215 ft²) should accommodate the average small-bore vertical magnet and console, but the position of the room is dictated by the need to ensure that the magnet's field will not be distorted by heavy metal objects (such as vehicles, trolleys, or engineering plant).

The radius of the sensitive area extends vertically as well as horizontally, depends upon the strength of the magnet and should be given at briefings. The magnet's field can interfere with heart pacemakers and electronic equipment.

A minimum height of 3 m (10 ft) is usually required to enable refilling of the liquid helium and nitrogen that cools the magnet.

Large-bore magnets such as whole-body instruments, in which a patient is introduced into a cylindrical horizontal magnet, require larger rooms and more stringent precautions but are not usually used in labs.

The MRI equipment will not operate satisfactorily in high temperatures so that cooling is needed in these rooms. The doors should be 1½ leaves wide.

2.2.9 Clean rooms

Clean rooms are usually associated with the manufacture of miniaturized components and pharmaceutical and medical products, but they are occasionally required in laboratory complexes. A clean room is an enclosed area that requires a lower level of airborne particulate contamination than normal and which would generally incorporate temperature and humidity control. These requirements are achieved by purging the room with air that has passed through a filtration and conditioning system. The rooms should be under positive air pressure to avoid the ingress of contamination.

The degree of cleanliness is based upon the number of particles within a certain size range found in an air sample. The particle measure used is a micron (μm) which is one thousandth of a millimetre (25 μm = 0.001 in). In the UK the standard for clean rooms is BS 5295, in the USA it is FS209. Technical supplement 5 contains information from BS 5295. Clean rooms are generally of modular construction and by specialists.

Ancillary rooms

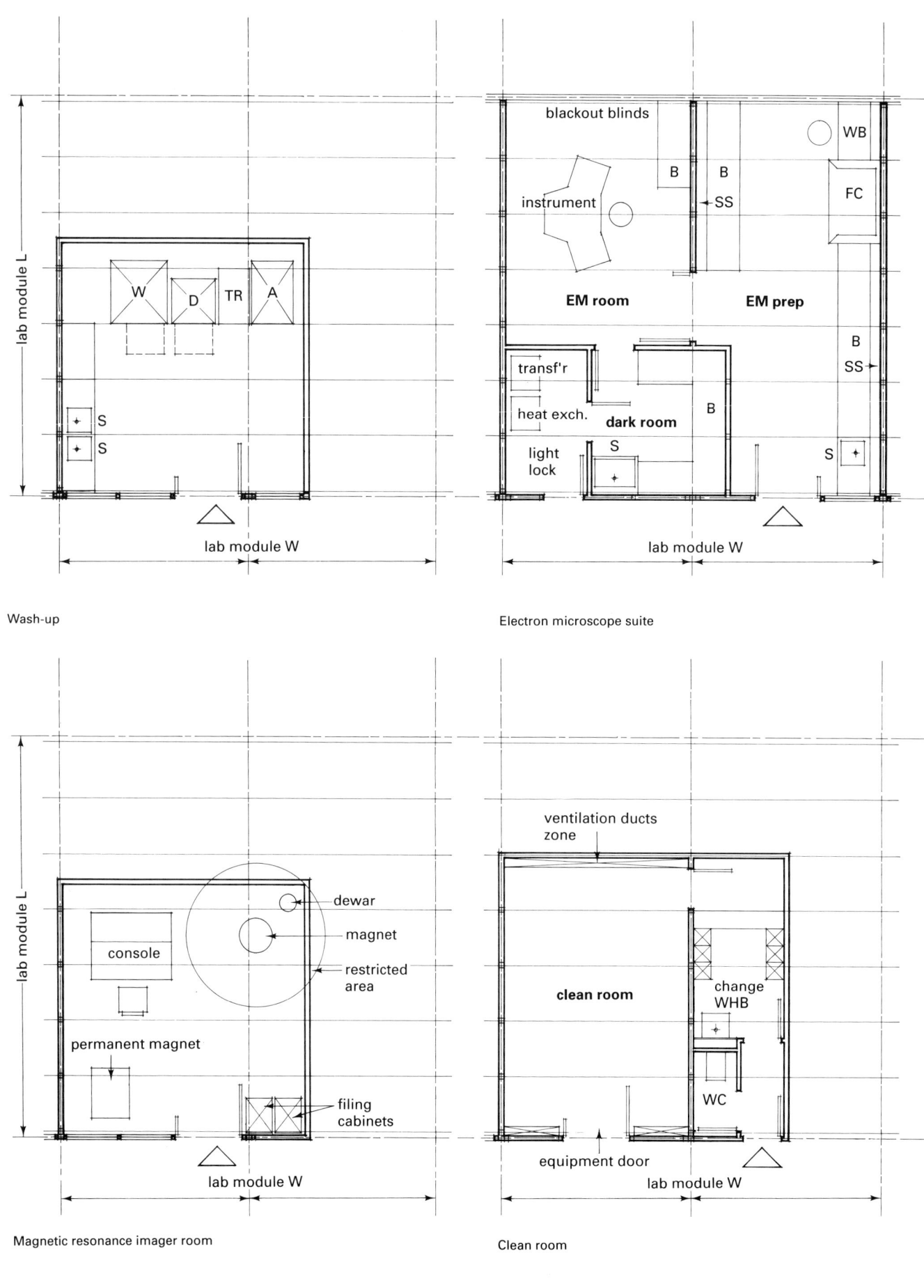

Wash-up

Electron microscope suite

Magnetic resonance imager room

Clean room

Laboratory spaces

2.2.10 Hydrogenation units

This accommodation is rarely encountered in lab projects and is very much tailored to the individual requirements of the person who will use it. It is therefore not sensible to provide specific information on the requirements, but the following general comments based on a built example may prove useful.

The unit consists of a room in which potentially very hazardous procedures using hydrogen are carried out with specialized equipment inside safety cabinets. Measures must be taken against explosions occurring within the room and to minimize damage outside the room should an explosion occur.

The room is best housed in a structure outside the lab building, but if inside the latter it should be on an external wall, with blast panels in that wall and in the ceiling if on the top floor. Partitions to adjoining rooms and the door, together with the structural floor and ceiling slabs, should be sufficiently strong to prevent an explosion from affecting adjoining spaces, resistance in the order of 490 kg/m^2 (100 lbf/ft^2) probably being required.

The room should have a small sink unit with hot and cold supplies, one or more unventilated safety cabinets and a safety bench. The latter should be 600 mm (2 ft) deep with the worktop 900 mm (3 ft) above the floor. The cabinet worktops should be the same height as the safety bench and their plan dimensions will depend upon the equipment to be used. An acceptable size for cabinets housing Parr 3911 equipment is 700 mm (2 ft 4 in) deep overall by 1100 mm (3 ft 7 in) wide overall. Both cabinets and bench should extend to the ceiling and be fronted with polycarbonate glazing, with fixed upper section and vertically sliding sash lower section, giving a clear opening of 600 mm (2 ft) high. In the example mentioned both cabinets and bench were provided by the fume cupboard manufacturer.

The cabinets require a supply of hydrogen, nitrogen and vacuum, together with socket outlets for a shaker motor, heater and cooler. Vacuum is also required to the bench, together with compressed air. Hydrogen, nitrogen and compressed air are probably best supplied from cylinders on wall brackets within the room, and vacuum by vacuum pumps powered by a water outlet over a dripcup mounted outside the cabinets and bench.

All electrics in the room should be flameproof.

The plan size of the built example was 3 m × 3 m (10 ft × 10 ft).

2.2.11 Computer rooms

With the miniaturization of hardware that has occurred in the past decade it is unlikely that the computer room in a laboratory building need be in excess of about 14 m^2 (150 ft^2), with a similar-sized *en suite* office for the person in charge.

2.2.12 Offices

Dependent upon the size and nature of the project these may be required for the director of a research unit, the head of each discipline, group leaders within the discipline, or the senior technician within the discipline. Offices will also be required for secretaries. The numbers and sizes should be given in the brief. Offices in lab buildings do not usually differ from those in any other building type.

2.2.13 Common rooms

These are meant for staff refreshments, as refreshments are not allowed within labs, and should be associated with beverage preparation facilities. In small units they may sometimes double up as a conference/seminar room. The size should be given at briefing.

2.2.14 Seminar room

Blackboard and projection screen should be provided and the room should be able to be blacked out and provided with mechanical ventilation.

2.2.15 Library

The main provision is for periodicals, including storage for back numbers. If a separate room is required then the size should be given in the briefing.

2.2.16 Copier rooms

To house photocopiers, faxes, etc. The use of these machines is increasing enormously, by both lab and office staff, and they should be separate from secretaries' offices and accessible to lab staff other than through secretaries' offices.

2.2.17 Plant rooms

The plant room is the heart of a laboratory building: it houses the engineering plant that provides the servicing and environmental requirements without which the labs cannot function. It ought therefore to be given a high priority by the design team.

Size

As a general rule there is a tendency at briefing stage to underestimate the amount of space that will be required for plant (virtually all of which requires access from all four sides) and it is almost unknown for a brief even to mention plant rooms except as a component of a notional 'balance area' allowance. The design team is likely therefore to look to the client in vain for guidance on plant room areas and the architect is unlikely to receive much help from the building services engineer on size until a decision has been taken on the positioning of the plant room(s), when the engineer will be able to make an initial assessment on the basis of 'what goes where'.

In the author's experience the plant rooms of research lab buildings almost always end up larger than the space initially allowed for them. With the inevitable absence of firm information at design concept stage it therefore behoves the designer always to keep in mind the need for a large plant room area (the traditional modest box on top of a flat roof will not do for these buildings). As the full extent of the area required will only be known at a relatively late stage, a loose-fit, open-ended and flexible allocation of space should be made. A pitched roof over the whole of the footprint of the building, with the eaves raised about 1500 mm (5 ft) above the floor slab, is one means of providing such a flexible area; it also means that all services to or from the labs emerge direct into the plant room.

The principle consumers of space in plant rooms are the air-handling units (AHUs), which supply fresh air to the building, and the extract fans, which extract the used air from the building and from the fume cupboards, together with their associated ductwork. Other items of plant housed are boilers, calorifiers, water treatment units, pumps, compressors, chillers (air-cooled condensers), water storage tanks, control panels, etc.

Position

The best position for fresh-air intakes (FAIs) is normally at the top of the building; fume cupboard exhausts must be taken up to discharge above roof level for maximum dispersion and room extracts are best discharged at high level for the same reason. Central air-handling plant is therefore most efficiently positioned above the top floor of the building and, all other things being equal, this is the best position for the plant room, with all plant in one location.

But much of the other plant – such as boilers, calorifiers, water treatment units and pumps – does not derive any performance benefit from being at the top level and, where space at that level is restricted, can just as well be sited at ground floor or basement level, with an adequate vertical duct connecting the two plant rooms.

Initial planning

It will not be possible to plan the layout of the plant room until a relatively late stage, when the plant requirements are known, but well before then it will be necessary to make certain 'in principle' decisions to enable the planning of the building to proceed. These are:

Plant room

Control panel

Boiler

Central heating pumps

Hot-water service pumps

Ancillary spaces

Plant room

Fume cupboard extract fan

Floor drain with equipment drains

Piped gases cylinder store

Gas cylinders in lab

- the position of the plant room(s) in the building (section 2.2.17b);
- the position of an unroofed area, preferably adjacent to the plant room and accessible from it, to house the chiller(s);
- the position of a zone for fresh-air intake louvres, along one or more sides of the plant room.

Layout

The internal plan form of the plant room is likely to be either a single corridor with plant on both sides (as Type 1 in Chapter 1, section 1.5.2) or two corridors with plant on the outside of each and in the space between (as Type 4) or variations of these, depending upon the size of the project.

The internal layout is usually planned by the building services engineer in collaboration with the architect. The following comments may be found to be useful.

- AHUs are essentially rectangular steel boxes made up of sections, starting from the inlet end, of filter, heater, cooler (if required), humidifier (if required) and fan, with ductwork from the inlet end to the FAI louvres and from the discharge end to the spaces to be supplied. Their size will depend upon the total volume of these spaces. Their most frequently accessed component is the filter, so this should be easily accessible.
- Extract fans will have ductwork from the rooms or fume cupboards served connected to the inlet end, and ductwork from their discharge end taken either horizontally to louvres (well away from FAI louvres) or vertically through the roof. Discharges from fume cupboards are always taken up vertically to above the roof.
- Boilers have a flue to be taken through the roof.
- The number of chillers required will determine the size of their open area. If possible, some spare area should be allowed for future additional chiller(s).
- Space must be provided for a central control panel, linked to each item of plant and indicating which is in operation. The panel, which should be at the entrance to the room, is usually about 600 mm (2 ft) deep and about 2 m (6 ft 6 in) high, may be 4–6 m (13 ft–20 ft) long, and must be accessible from both back and front. Mains power to the panel will be cabled from the main switchboard (Chapter 5, section 5.2.1).
- Space must be allowed at high level for water storage tanks.
- Cabling between the control panel and each item of equipment is usually carried on cable trays at doorhead height, with drops down to the item to be served.

2.3 Storage areas

2.3.1 Central stores

For bulk storage of equipment and apparatus. The size, together with information on the items to be stored, should be given in the brief. The area may be internal and should be convenient to the supplies and goods reception area. The door should be a minimum 1½ leaves wide.

2.3.2 Solvents/inflammable liquids stores

Usually entirely separate from the main laboratory building, with fire-resisting doors and permanent ventilation, and with blow-out panels to dissipate the force of an accidental explosion, preferably in the roof (light-sheeted roofs provide this facility). In the UK proprietary prefabricated buildings are available specifically for such stores.

2.3.3 Radioactive stores

May be required for holding radioactive waste material until it is collected for disposal. In most research labs radioactive material (such as isotopes) is brought in and circulated in lead-lined containers, so that no separate store is required for the unused material.

2.3.4 Special gases cylinder stores

These may consist of a central bank of cylinders from which the special gases are piped to outlets in the labs, or a store from which cylinders of the special gases are taken into the labs for use, or a store for storing the empty cylinders.

The location of the stores will be dependent upon the agreement of the fire officer; they are normally outside the main building and, where the last two are concerned, are often merely meshed unroofed enclosures with lockable gates. In the first, piped gas installations, it is usual for two cylinders of each special gas to be connected by a manifold from which the piping to the labs is taken; this enables the gas supply to be switched from the empty to the full cylinder and for the former to be replaced without the supply to the labs being interrupted. Prefabricated modular gas houses of epoxy-coated steel are available with mesh doors and complete with all the necessary fittings including manifolds, a suitable type being supplied by Mason Nordia Ltd in the UK. Storage of connected acetylene cylinders is subject to extra restrictions: in the UK the facility must have at least 25% of the surface area open to the atmosphere, must be separated by an impermeable partition of 1 hour fire resistance from any working space within the building to which it is attached or which is within 8 m of it. Any window or door within 3 m of the facility must have a fire resistance of ½ hour.

2.3.5 Chemicals store

For the bulk storage of chemicals: may be required for chemical labs, adjacent to the labs. Should have permanent ventilation to prevent the build-up of vapour, and if internal should have extract ventilation.

2.4 Workshops

The provision of workshops will depend upon the size and complexity of the laboratory complex, and should be specified in the brief if required.

2.4.1 Heavy workshop

Will house floor-mounted machines such as lathes, grinders, millers and saws, some of which will require three-phase power and compressed air. The rooms should contain a sink unit with hot and cold water, and have some workshop benching with power, fuel gas and compressed air outlets, together with provision for materials storage.

2.4.2 Instrument and electronic workshops

For delicate or precision work and the repair and maintenance of instruments and electronic equipment. The rooms should contain a sink unit with hot and cold water. Much of the work is carried out on benching similar to lab benching, with similar provision for power, fuel gas and compressed air.

2.4.3 Other workshops

More specialized processes such as welding and paint spraying may require separate rooms or spaces.

2.5 Animal areas

2.5.1 Introduction

Laboratory animal houses are a building type in their own right and will only be covered in general terms in this book (refer to the Further reading section for additional information). Like labs, animal houses may vary in size between a small number of rooms for one species with basic ancillaries, to a multi-storey building catering for a variety of species with sophisticated back-up facilities.

Animals are used in a wide range of laboratory experiments: they may be used for toxicological assays on groups of animals treated with drugs, sera, vaccines, insecticides, hormones, radiations, cancers etc.; for behavioural, genetic, nutritional and metabolic studies; for diagnosis of disease in man; for surgical experiments and organ transplants; or for studies of particular diseases.

The type of study will determine the species to be used; the most common are small rodents, with mice being extensively used. Larger animals such as cats, dogs and primates are used for certain kinds of research but their accommodation is more sophisticated.

To be of use animals must be of the right weight and age, free of any conditions that could affect the results of the experiments being conducted and in the right numbers. Special-pathogen-free (SPF) animals call for the provision of particularly stringent conditions. To ensure that these parameters are met it is necessary for animals to be kept in very carefully controlled environmental conditions and to be isolated from sources of contamination and disturbance.

Animal houses are therefore usually self-contained units, isolated from other accommodation and with access restricted to designated, licensed staff. They should be near the laboratories in which they are to be used but in separate buildings, wings or floors of the lab building.

In the UK animal house accommodation and procedures are tightly controlled by the Home Office: both the accommodation and the staff caring for the animals in that accommodation are required by law to be licensed by the Home Office.

The legal protection of lab animals is covered in the UK by the Cruelty to Animals Act 1876, and in addition they are protected by the Protection of Animals Act 1911, the Protection of Animals (Anaesthetics) Acts of 1954 and 1964, and the Veterinary Surgeons Acts of 1948 and 1966.

In 1986 the Animals (Scientific Procedures) Act was introduced in the UK and, pursuant to that Act, in 1989 a Code of Practice for the Housing and Care of Animals used in Scientific Procedures became law. The latter is the publication of most relevance to designers, containing guidance on the physical and environmental requirements. It sets out basic principles. Plans of proposed new buildings should always be submitted to the relevant controlling authority for comment at a preliminary stage.

Opposition to the use of animals in experiments is increasing and becoming more militant. Extremist groups have attacked animal houses and in some instances those working in them, so that security is now a very high priority.

A good brief is especially important in animal houses, incorporating the requirements of the scientists and animal technicians, who should be familiar with the ramifications of the statutory requirements and should always be consulted by the designer on these.

2.5.2 Animal house accommodation

Animal rooms

Sizes should be given at briefing, together with numbers required. As a guide, a minimum width of 3 m (10 ft) is usually adequate with a length from 4.5 m to 6 m (15–20 ft). Smooth, washable, impervious surfaces are required to walls, ceilings and floors, the latter usually falling to a floor drain.

Because of the environmental requirements it is usual for natural light to be excluded. Door leaves should be minimum 1050 mm wide × 2100 mm high (3 ft 6 in × 7 ft), with an observation panel that can be blacked out.

Cage washing

Should be divided into clean and soiled areas, with sink and sterilizers or autoclaves between.

Stores

Required for food, bedding and cages.

Animal houses

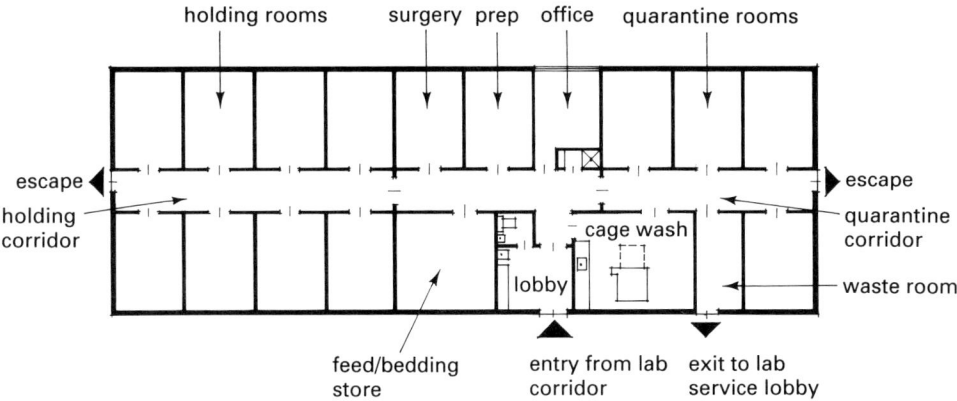

Animal house plan

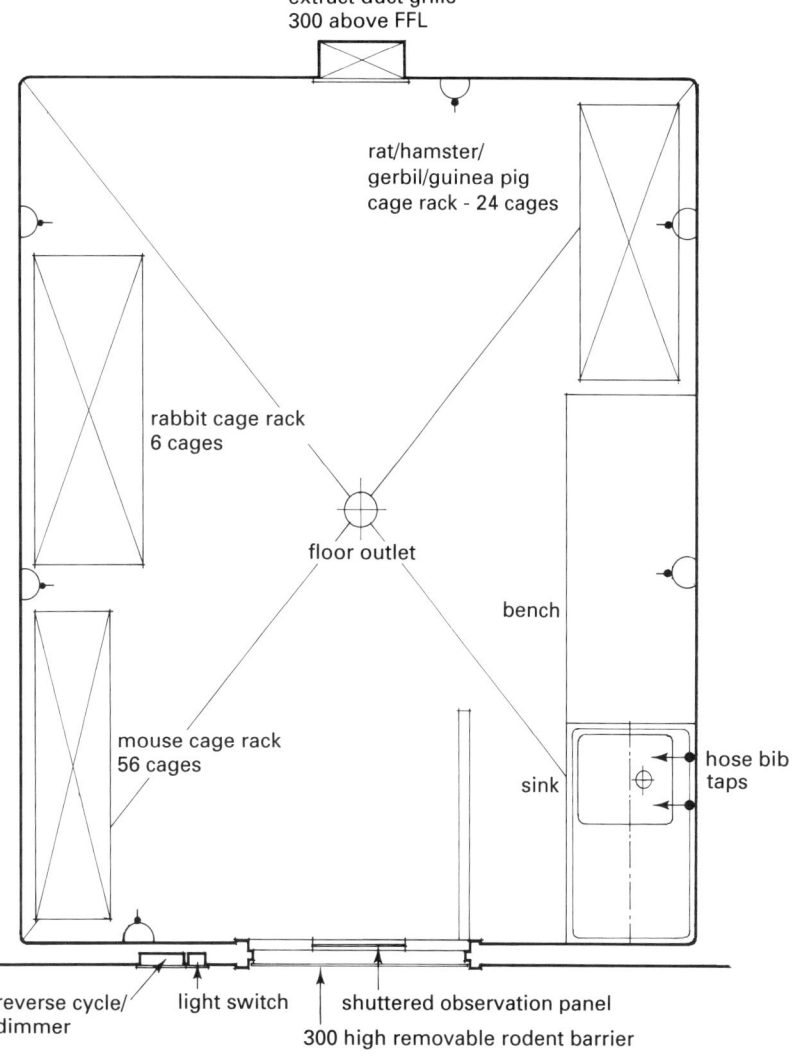

Rodents room plan

Laboratory spaces

	Staff changing and common rooms	Should be arranged as an entrance lock to the animal house, usually with lockers, toilets and showers. Personnel should only proceed beyond this point when wearing gowns and overshoes. If possible, common rooms should have natural lighting.
	Offices	Should have natural lighting and be in a position where they control access to the animal house.
	Ancillary rooms	Should be planned in close proximity to the animal rooms but separated from them by a cut-off door and may include preparation rooms, surgeries, post-mortem rooms, X-ray rooms, etc.
	Refuse rooms	For holding soiled bedding, waste, etc., prior to disposal and may house incinerator. Refuse should be placed in sealed bags in the animal rooms before removal to the refuse room.
	Circulation	In SPF units separate clean and contaminated corridors are required, but in standard animal units a single corridor will suffice, usually 1800 m wide (6 ft).
2.5.3	Fixtures	
	Cages	Cages may be supported on shelving fixed to the walls, but it is now more usual for caging to be supported on mobile racking independent of the walls. In the UK a number of firms such as NKP Cages and Forth-Tech Services, produce racking with the cages to fit into it, together with separate floor-standing cages for larger animals. See Chapter 4, section 4.2.18 for typical cage sizes.
	Fittings	A small sink unit is usually provided in animal rooms, with shelving over, and movable benching may be required.
2.5.4	Services	Hot and cold water are required to the sinks in animal rooms together with a mains water supply for drinking water. Waterproof socket outlets are required in all animal spaces. Steam will be required in cage-washing rooms, and other services that may be required should be identified at briefing.
2.5.5	Environment	Full air conditioning (mechanical ventilation with heating, cooling and humidity control) is usual in animal areas, with no recirculation of air. In the UK the recommended environmental conditions for the various species are given in the Code of Practice for the Housing and Care of Animals. Animal areas have a distinctive smell and technicians work in the enclosed environment for long periods. Natural ventilation with a view out should therefore be provided in offices and common rooms wherever possible.
		Partitions should provide a sound reduction of at least 45 dB, as small animals are extremely sensitive to noise and experiments can be spoiled if the animals are disturbed.
		Lighting in animal rooms should be controlled by dimmer, from outside the room, to enable different daylight conditions to be simulated.
2.6	**Circulation**	
2.6.1	Corridors	Should be a minimum 1500 mm (5 ft) wide, but preferably 1800–2000 mm (6–6 ft 6 in).
2.6.2	Lifts (elevators)	Should be goods/passenger, large enough to carry lab equipment, i.e. minimum car size 1700 mm × 2300 mm (5 ft 6 in × 7 ft), minimum load capacity 1000 kg (2200 lb).
2.6.3	Space	Between benchtops or equipment should be as illustrated.

Support system

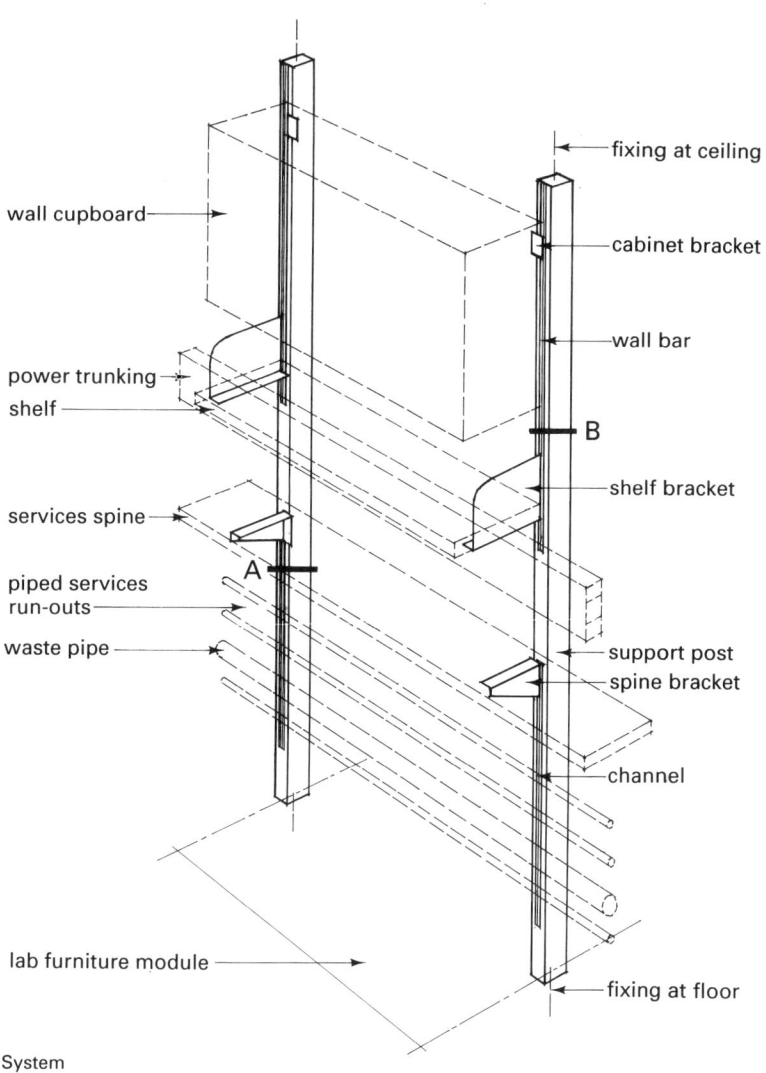

System

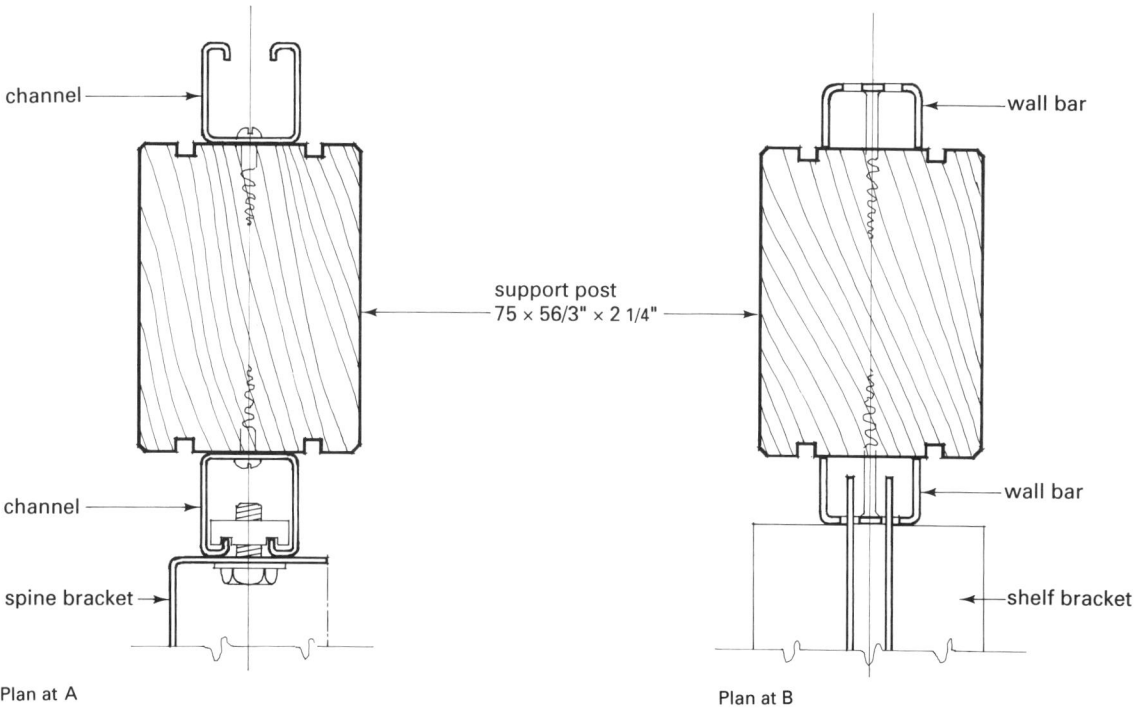

Plan at A

Plan at B

Fitting out laboratory spaces

3 Fitting out laboratory spaces

3.1 Support systems

Support is required for services, spines and bridges, for shelves and wall cupboards, for power trunking, for the piped services run-outs supplying the outlets on the services spines, and for waste pipes.

The most flexible means of supporting shelves and wall cupboards is by means of proprietary slotted wall bars into which brackets can be hooked to carry shelves, and cabinet brackets can be hooked to support wall cupboards. With this system shelves and cupboards can be raised or lowered, removed or added or swapped over. One suitable system is supplied by Spur Shelving.

Services ledges, piped services run-outs and waste pipes are best supported on proprietary channels with captive nuts, into which brackets and pipe clips can be fixed at exactly the required height. One suitable system is supplied by Unistrut, their M5 small channel being adequate.

If brick or block partitions are to be used then the wall bars/channels and power trunking can be plugged and screwed direct to them.

A more suitable type of partitioning for labs is a post-and-panel system, with the posts at the same centres as the lab module and carrying the wall bars and channels and power trunking. Panels may be solid, glazed or omitted altogether (as in peninsular benching in labs consisting of a number of lab bays) as required, and can be changed as the function of lab bays change. This system also provides space within the partition thickness for the pipe and conduit vertical drops that supply the run-outs and power trunking in installations where the supply is from above.

Suitable post-and-panel systems are marketed by some lab furniture suppliers but these require the use of the supplier's furniture, thus excluding competitive tenders for the lab furniture and requiring the supplier's presence on site earlier than would otherwise be the case. In practice, suitable systems can be designed in timber for individual projects without incurring extra costs and enabling the partitioning to be carried out by the main contractor: an advantage in terms of programming (and usually cost) in the UK. One such system is described in Chapter 6, section 6.2.

Several lab furniture manufacturers supply fully supported and enclosed services spine and services bridge units, but these do not usually provide high-level support for shelves or wall cupboards. The units normally include all run-outs, services outlets, wiring and power sockets and, although very neat, are less flexible than the system described above and require the manufacturer's presence on site earlier than would otherwise be the case.

Services spines

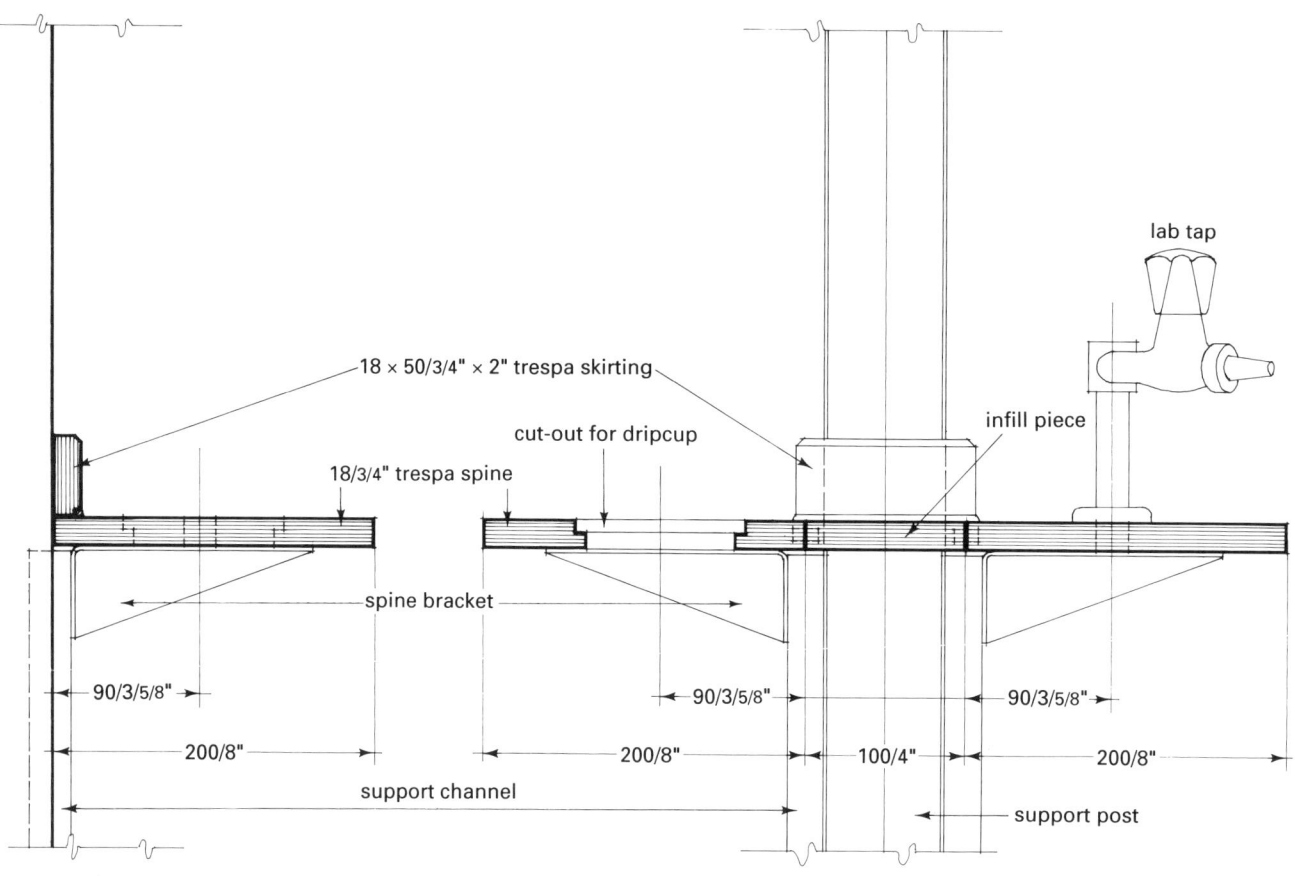

Wall spine section

Peninsular spine section

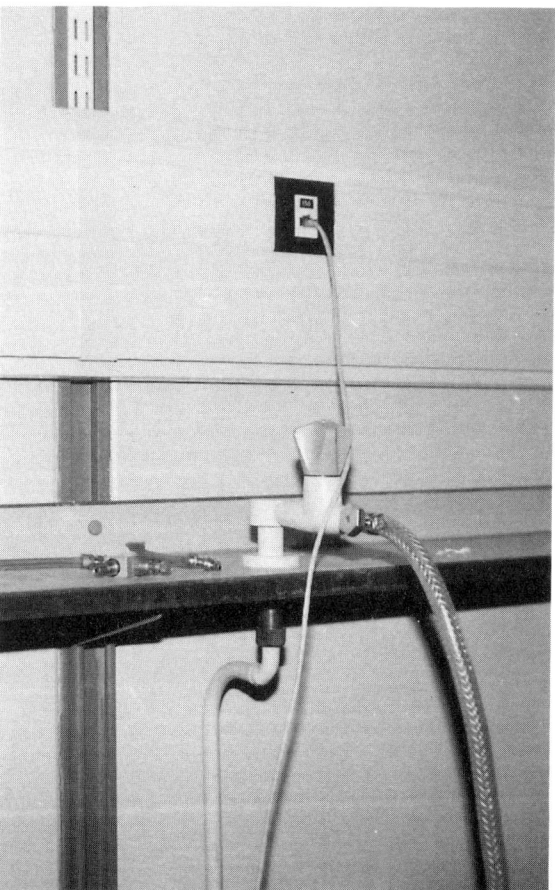

Wall spine

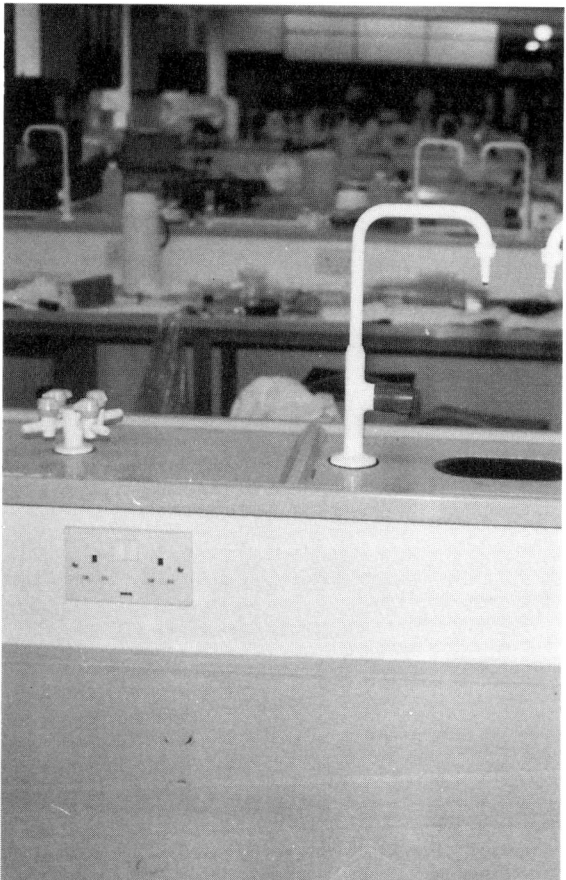

Raised peninsular spine

Fitting out laboratory spaces

Services bridges

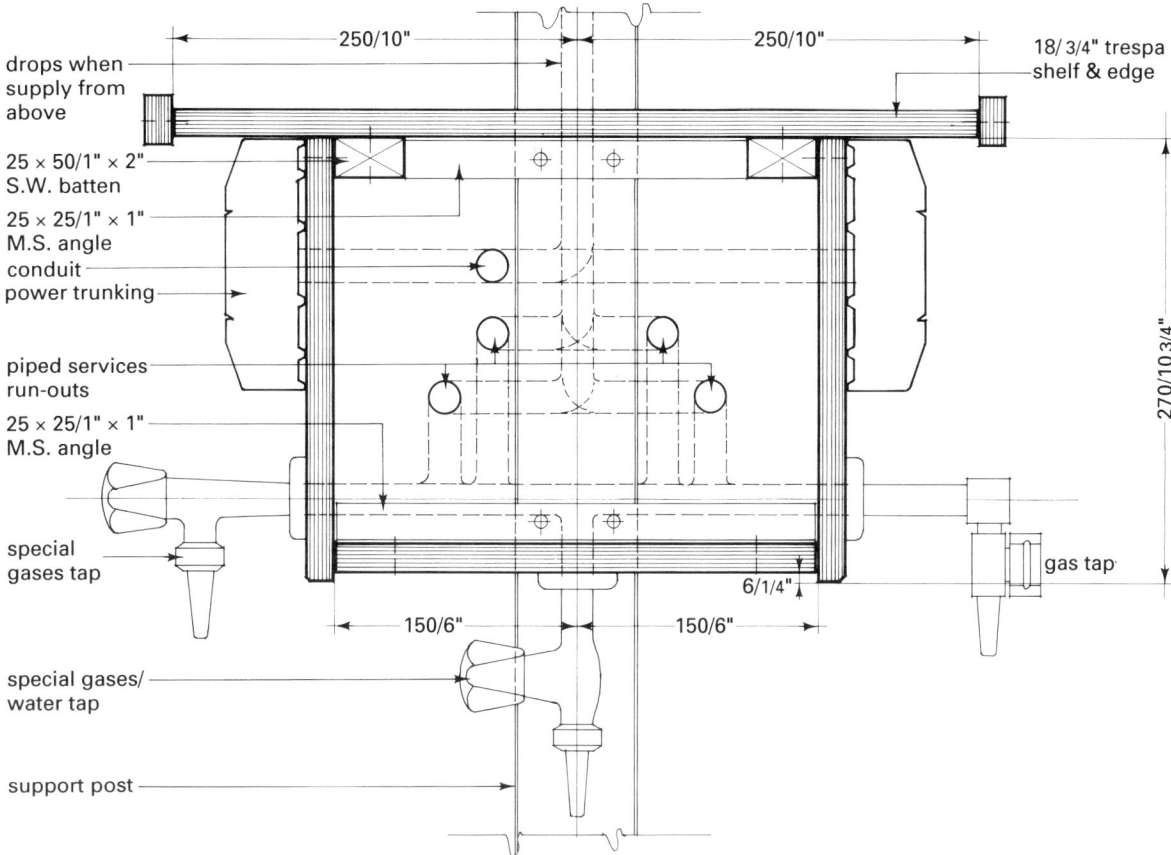

drops when supply from above
250/10"
250/10"
18/3/4" trespa shelf & edge
25 × 50/1" × 2" S.W. batten
25 × 25/1" × 1" M.S. angle conduit power trunking
piped services run-outs
25 × 25/1" × 1" M.S. angle
270/10 3/4"
special gases tap
6/1/4"
gas tap
150/6"
150/6"
special gases/ water tap
support post

Section

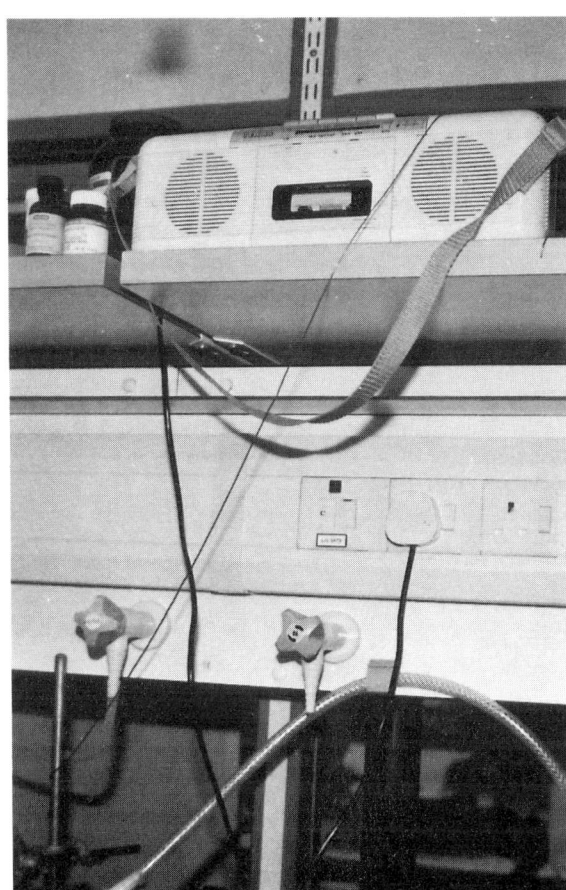

Bridge close-up

Bridge in position

Services spines and bridges

The benefits of restricting the work carried out by the lab furniture supplier to lab furniture components only are contractual within the UK contract scene, and may not be valid in other countries.

3.2 Services spines and bridges

The old practice of mounting bench outlets on benchtops was inherently inflexible because it anchored benches to services, and it has generally been discontinued except in teaching labs. Bench services outlets are now normally mounted separately from the benches, either in spines behind the benches or on bridges above the benches.

Services spines are fixed ledges (or shelves) on which the bench-type services outlets and dripcups are mounted and against which benching is set. Spines require a depth of 200 mm (8 in) from back to front to accommodate the dripcups (tundishes) together with the piped services run-outs below with their connections up to the outlets in the spine. Spines normally run the full length of lab partitions or the supports of peninsula or island benches, but stop either side of fume cupboards to allow these to be positioned hard back against the partitions (the piped services run-outs and the wastes continue through uninterrupted behind the fume cupboard supporting the underframe/cupboard). Spines are usually level with benchtops, with the latter butting against them, to form an extension to the benchtop, with outlets projecting up through holes in the spine and dripcups recessed flush with the spine. They may also be about 200 mm (8 in) above bench height, with outlets and dripcups as above but with a fascia between spine and benchtop to accommodate power outlets. Spines are supported as described in the previous section and are usually made of a material similar to the benchtop.

Services bridges are mounted at a height above the benchtop sufficient to enable equipment on the bench to continue through under them, and normally only occur in peninsula or island bench locations. They consist of sides, bottom and top, enclosing the piped services run-outs and electrical conduit, with power outlets in the sides and services in the sides or bottom, the top serving as a shelf. If dripcups are required then these will need to be in a service spine separate from the bridge at bench level. The space available for piped services run-outs in the bridge is necessarily restricted and is shared with power conduits, so that the range of services that can be provided is less than with the spine system and their installation and modification is that much more difficult. Services bridges are not usually used in labs with an extensive range of piped services.

Bridges require a minimum depth of 300 mm (1 ft) to accommodate the run-outs together with bends and connections to outlets, and a similar minimum height to accommodate services and power outlets in the sides. This height may need to be increased if the range of piped services provided is more than two or three.

Bridges will be supported as described in the previous section and will usually be constructed of solid laminate, or medium-density fibreboard (MDF).

The separation of services spines and bridges from the benching leaves the designer the choice of whether or not they are to be by the specialist lab furniture supplier or by the main contractor. The delivery times for lab furniture can be considerable and with some refurbishment projects these times can actually dictate the contract period. Any work that does not necessarily have to be carried out by the lab furniture specialist, which involves other trades and is required a considerable time before the lab furniture proper, may therefore benefit by being left to the main contractor.

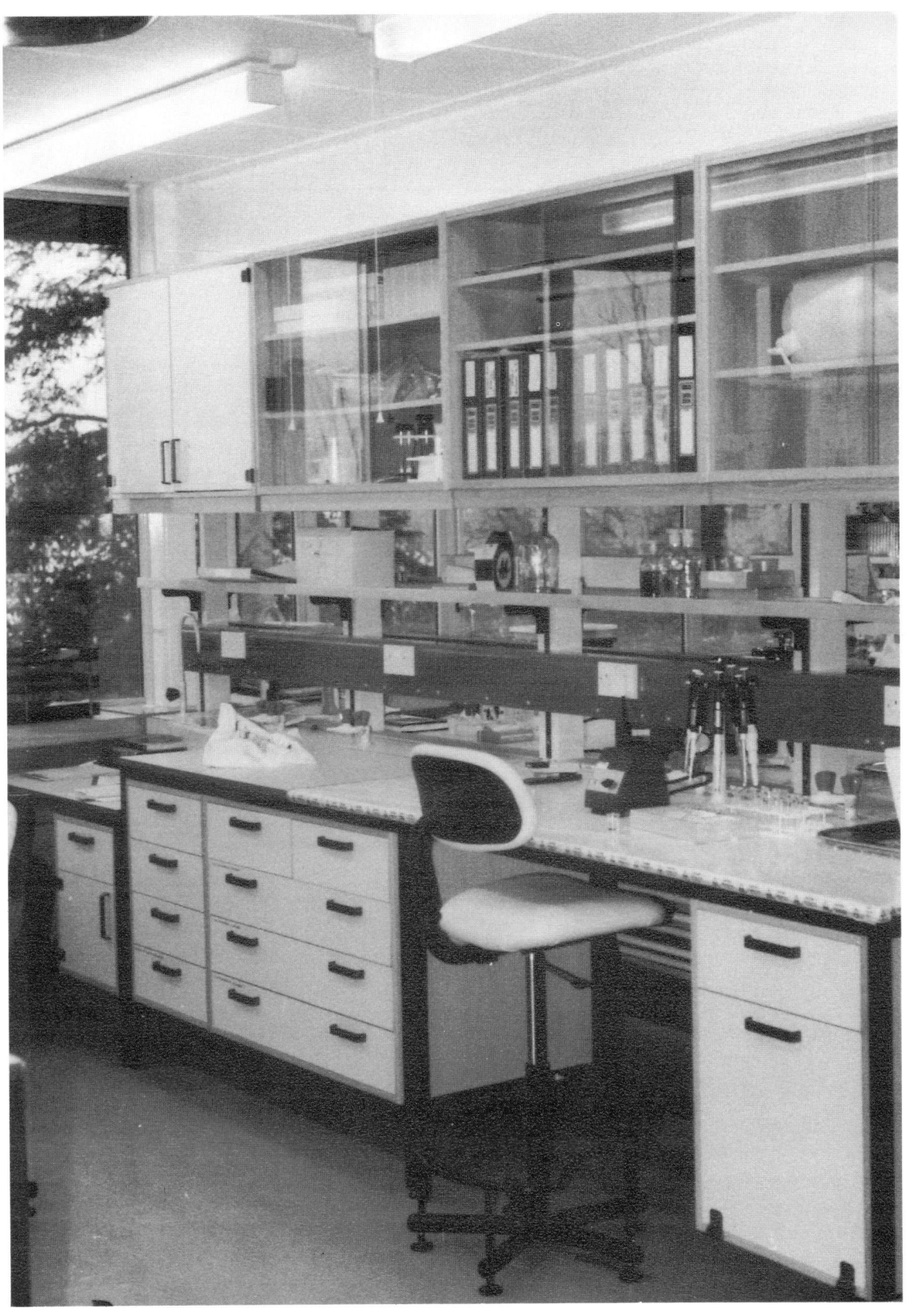

3.3 Laboratory furniture

3.3.1 Introduction

In the UK Laboratory Furniture and Fittings are the subject of BS 3202 : 1991. Lab furniture is usually supplied and fixed by specialist manufacturers, of whom there are a relatively large number, all making broadly similar products. All offer their own standard ranges but most will quote for and supply variations to these, to suit individual projects, and most subcontract the fixing to specialist teams.

Lab furniture for all but the smallest and simplest projects will be manufactured for the particular project, even if chosen from the manufacturer's catalogued range, and delivery times can be in the order of 10 to 12 weeks, which can be critical on some refurbishment projects.

In the past, with services outlets mounted on benches and piped services run-outs incorporated within the bench structure, it was customary for lab furniture suppliers to be responsible for supplying and fixing the outlets and services run-outs, as well as the furniture itself. This could occasion problems of divided responsibility between the furniture supplier and the mechanical and electrical subcontractors and, in today's contractual climate, is to be avoided.

The separation of services from benches into spines or bridges, as described in the previous section, obviates the need for the lab furniture supplier to become involved in such non-furniture work. In the author's experience the supplier should be set to concentrate on the furniture proper – benches, cupboards and shelving – which, because it is not serviced, can come on site very late in the contract.

Most manufacturers offer a modular range of components. UK lab furniture suppliers tend to favour the 1000 mm (3 ft 3 in) module, while other European suppliers favour the 900 mm and 1200 mm (3 ft and 4 ft) modules – and one even an 800 mm (2 ft 8 in) module!

The choice of module should be made at an early stage of the project, as it should determine the length of the lab module (Chapter 2, section 2.1.4) and will determine the support system centres (section 3.1). The author has used the 1000 mm module for over 10 years and has found it to be convenient; it also accords with standard metric sink unit sizes and with fume cupboard sizes. UK manufacturers will all supply to the 1000 mm module and other European firms may do so on large projects; it seems probable that in future most firms will supply to whatever module is chosen, although this should be checked before tenders are invited, especially if the client has a favourite supplier whose 'house' module is not that chosen by the designer.

The usual procedure for handling the lab furniture is to submit to the client a 'menu' (see Technical supplement 9 for an example) offering a range of configurations for each component, and to invite competitive tenders for these on the basis of a performance specification (see Technical supplement 8 for an example). This enables each tenderer to offer standard components where these are acceptable and includes detailed specifications only when these are a definite requirement (for benchtop material, for example). The main lab furniture components are benches, underbench units, wall cupboards, shelving and sinks. See Technical supplement 10 for information on lab furniture suppliers.

3.3.2 Benches

The removal of services from benches means that these can become movable, to be rearranged in position or removed completely to make way for equipment, and this has virtually become the norm. The most common system is for benching to consist of 'table-benches' – benchtops mounted on tubular steel underframes – which are pushed into position against the

Benchtop materials

Material	Resistance to: Chemical attack	Staining	Heat	Solvents	Absorption	Indentation	Impact	Flatness
Teak/iroko	Fair	Fair	Fair	Fair	Fair	Fair	Good	Fair
Laminate	Fair	Fair	Poor	Excellent	Excellent	Good	Good	Good
Solid laminate	Fair	Fair	Fair	Excellent	Excellent	Excellent	Excellent	Excellent
Epoxy resin	Excellent	Excellent	Good	Excellent	Excellent	Excellent	Excellent	Good
Ceramic	Excellent	Excellent	Good	Excellent	Excellent	Excellent	Excellent	Good
Tile	Excellent	Good	Excellent	Good	Good	Excellent	Good	Fair
Stainless steel	Good	Good	Excellent	Excellent	Excellent	Good	Good	Good
Asbestos cement	Good	Poor	Excellent	Poor	Poor	Good	Excellent	Good
Slate	Good	Good	Good	Good	Fair	Excellent	Good	Excellent
Polypropylene	Good	Good	Poor	Fair	Good	Good	Good	Fair

Laboratory furniture

Bench systems

Criteria	Pedestal*	Table bench†	Cantilever††
Clear floor under units	No	Yes	Yes
Underbench units easily removed	No	Yes	Yes
Bench rigidity	Excellent	High	Fair
Bench loading	High	High	Good
Access to service pipework	Poor	Good	Good
Bench movability	Poor	Excellent	Poor
Cost	Low	Average	Average

* Bench supported on underbench units.
† Bench supported on underframe, underbench units suspended from frame.
†† Bench supported on cantilever frame, underbench units suspended from frame.

Writing benches

services spines and are modular, with removable underbench units either suspended from the underframe or floor-standing. An alternative system employs cantilever underframes secured back to the wall, with underbench units as before, but this is less flexible.

The most common benchtop depth is 600 mm (2 ft), with a nominal height of 900 mm (3 ft) for workbenches and 750 mm (2 ft 6 in) for writing benches, but other depths and heights can be specified with little, if any, cost implications. 800 mm (2 ft 8 in) depths are sometimes required to accommodate bulky bench-mounted equipment, and the need for these should be established in the brief.

Benchtops may also be in long lengths and supported on underbench units, but this is becoming increasingly rare because of the lack of flexibility. It is usual for benchtops and underframes to be in multiples of the furniture module (1, 1½ or 2 modules long). This enables benches to be removed and replaced with a similar module writing bench, sink unit, or fume cupboard.

The type of bench system to be used on the project should be agreed with the client when preliminary proposals are made, as should the module size.

Underframe legs should be square steel tube, minimum 30 mm, with bottoms fitted with adjustable feet for levelling.

Benchtops are available in a wide range of materials, including solid laminate, laminate on chipboard core, PVC/polypropylene/polythene on chipboard core, hardwood, stainless steel, and epoxy resin. All suit particular conditions, as shown in the table on page 43. The most commonly used for research work is solid laminate (Trespa or Print) 18 mm or 20 mm (¾ in) thick, which has all the benefits of laminate on chipboard core without the drawback of having to lip the edges, as the core is homogeneous with the surface material.

For applications such as radioactive work, where a joint-free raised edge is required around the edges of the worktop to contain spillage of radioactive material, epoxy resin is the most suitable (Simmons or Durcon).

Most benchtop materials, with the exception of solid timber, come in maximum lengths of just over 3 m (10 ft). For most applications the joint between two table-benches pushed tightly together will be acceptable, but this should be confirmed by the client. If this is not acceptable, and a crack-free sealed joint is required, then it will be necessary to use epoxy resin (which can be flush-jointed with epoxy resin cement), or a material that can be welded, such as stainless steel, PVC, polypropylene or polythene. No completely satisfactory method of achieving a smooth, sealed flush joint to solid laminate or laminate on chipboard core has to the author's knowledge yet been discovered. Where sealed joints are required between these worktops, and between the worktops and the services spines, a silicone sealant is usually applied.

Balance benches, which are basically table-benches but of heavier construction and with vibration dampers, are available from some manufacturers, in the same module as their lab furniture.

3.3.3 Underbench units

These may be pedestal units, which have recessed skirting/plinths and stand on the floor, or suspended units, which hang from the bench underframe with their bottom about 150 mm (6 in) above floor level. The latter may be slid sideways along the underframe rails or unclipped and removed completely. The former can be moved in position under the bench or removed completely, unless they are being used to support the benchtop.

Underbench units will be ½ module or 1 module wide, the widths such that two ½ modules or 1 module will fit between the legs of a 1 module bench.

Lab furniture

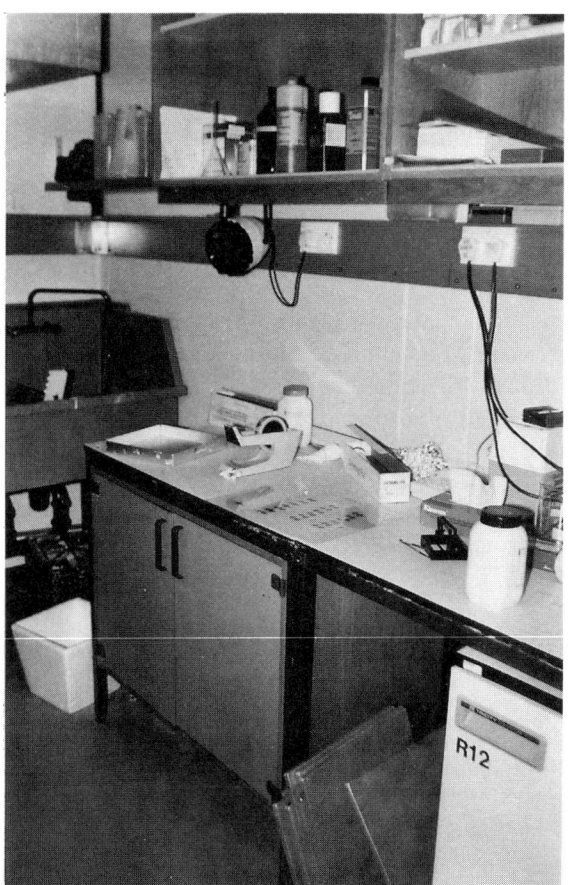
Underbench cupboard unit

Underbench drawer unit

Wall cupboard: glass doors

Wall cupboard: solid doors

The units are available as cupboard or drawer units, or cupboard-and-drawer units, in a number of arrangements and also in HFL (highly flammable liquids), ASC (acid storage) and waste bin versions.

Unit carcasses are usually in medium-density fibreboard or chipboard, the former finished with acid-catalysed lacquer or laminate or veneer, the latter with laminate or veneer. If a particular finish is required this should be specified. Doors may be hung within the frame provided by the carcass or 'laid on' to cover the carcass completely. The latter is less expensive and less fussy-looking and is usual nowadays, using special hinges that enable the door to open to 90° without projecting beyond the side of the carcass and which have a self-closing action to keep the door closed (as is common in kitchen cabinets). Handles vary, and any preferences should be specified. Doors are usually of chipboard or MDF, laminate-covered. Drawer fronts will be of similar construction and finish to the doors.

Cupboard units are normally fitted with one adjustable shelf.

3.3.4 Wall bar systems

These are proprietary slotted channels about 25 mm × 16 mm (1 in × ⅜ in), with brackets hooked into the slots to support shelves and wall cupboards.

Some lab furniture suppliers have their own in-house systems, but it is better to specify a system that is widely available through builders' merchants so that, when additional shelves are required, the brackets and any additional wall bars can be purchased without recourse to the lab furniture supplier. In the UK, Spur Shelving supply a suitable system.

In order to ensure that the wall bar centres exactly match the lab module, they should be supplied and fixed by the lab supplier.

3.3.5 Wall cupboards

These are generally 1 module wide, 300–350 mm (12–14 in) deep, in two heights approximately 600 and 900 mm (2 and 3 ft) high, with one adjustable shelf for the former and two for the latter, and fitted with sliding glass doors or solid doors. Different firms have slight variations in the depths and heights that they offer, but these are not usually critical.

Carcasses of wall cupboards will match those of the underbench units, and solid doors, with hinges and handles, will be similar to the underbench units.

Sliding doors should be 6 mm (¼ in) polished glass in PVC channels, with finger pulls.

The carcass should be fitted with a downstanding rail along the front edge of the bottom to prevent sagging.

Wall cupboards should be hung from the wall bar supports using cabinet brackets supplied by the wall bar system supplier, which are screwed to the backs of the cupboards. Alternatively, where there are no wall bars, the cupboards may be screw-fixed through the back to the partition.

3.3.6 Shelves

These should be in 1 module widths and are best supported on bookshelf-type brackets slotted into the wall bars. Reagent shelves are usually 225–250 mm (9–10 in) deep and with upstand lip to front edge; storage shelves are generally 300 mm (1 ft) deep but may be of greater depth – 450 mm (1 ft 6 in) or 600 mm (2 ft) – if required. Reagent shelves may be of solid laminate or laminate-faced blockboard; storage shelves are normally the latter. Blockboard 25 mm (1 in) thick should be used to avoid sagging.

3.3.7 Sinks

These may be the inset type or the sink-unit type. The former consists of a sink bowl inset into a benchtop for use as a source of water supply or running to waste, where greater capacity than a dripcup is required but

Lab furniture

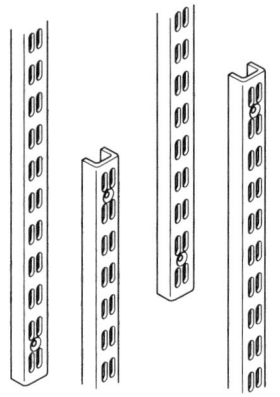

Wall bars

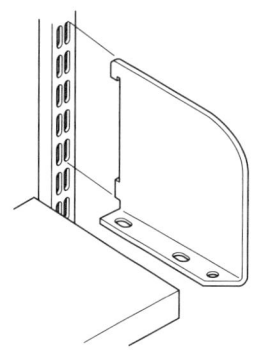

Shelf bracket

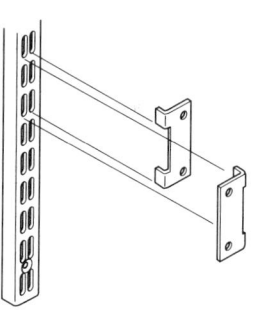

Wall cupboard brackets

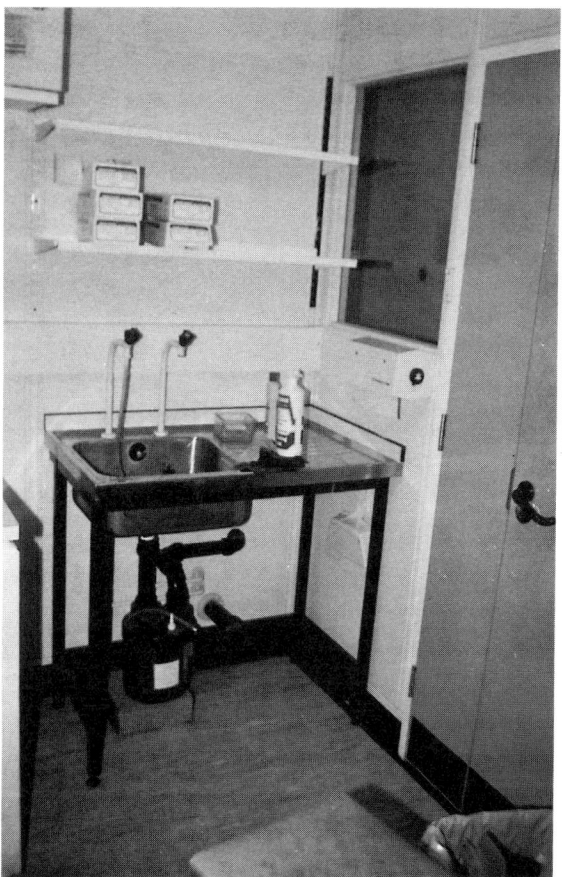

Sink unit

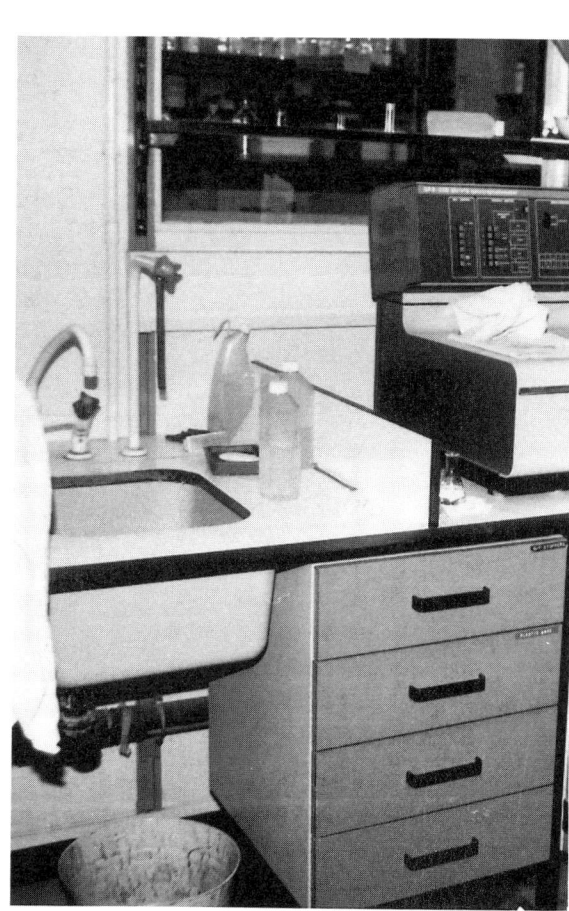

Inset sink, flange under

Fitting out laboratory spaces

where washing up will not usually take place. In the sink-unit type the bowl(s) and adjoining draining board(s) are formed from one material, without joints, for washing up.

Inset bowls generally occur in benchtops of a material different from the bowl, when it is not possible to weld the two together to form a seamless joint. The flange on the rim of the bowl must either sit on top of the worktop or be fixed against the underside, in each case with a sealant between flange and top. The flange on top gives a cleaner installation and does not require a finished edge to the cut-out in the top, but spillages are less easily swept into the bowl. The client's preference should decide which to adopt. Bowls are available in various materials, including stainless steel, polypropylene and epoxy resin.

Sink units are bowl-and-drainer combinations mounted on bench-type underframes and usually match bench sizes, i.e. 600 mm (2 ft) deep, 900 mm (3 ft) high and 1, 1½ or 2 modules long, so that they can be interchanged with benches. They may be single bowl/single drainer (left-hand or right-hand drainer), single bowl/double drainer or double bowl/double drainer, and normally have tapholes in the sinktop. Sink units are usually in stainless steel but are also available in epoxy resin.

The stainless steel grade used in laboratory sinks should be the acid-resisting 316 S16 grade, not the standard 314 domestic/catering grade.

Inset sinks normally arrive ready fixed in the benchtop, supplied by the lab furniture supplier.

Because 316 grade sinks are only made to order, all laboratory sinks are specials and likely to be on long delivery. There is therefore nothing to be lost and much to be gained by including the sink units, as well as the inset sinks, in the lab furniture subcontract, when they will be on the same module and their underframes will match the bench underframes.

3.4 Fume cupboards and hoods

3.4.1 Fume cupboards (fume hoods or laboratory hoods in the USA)

A fume cupboard is defined as a partially enclosed workplace that limits the spread of toxic fumes to operators and other personnel, is ventilated by an induced flow of air through an adjustable working aperture, dilutes the fumes and by means of an extract system provides for their release remotely and safely to the atmosphere.

Chemical and biochemical disciplines make most frequent use of fume cupboards, but some provision should be made in most laboratories for flexibility and possible future use. The fume cupboard standard in the UK is BS 7258 : 1990, but one fume cupboard manufacturer says that this BS is merely a badly presented rehash of DD 80 : 1982 in which it is difficult to find anything, that it is unsatisfactory and is likely to be superseded by a European Union Standard in due course.

A fume cupboard is a ceiling-height cabinet with an openable glass front and a work surface at bench level, supported upon a steel underframe or underbench unit, and with an outlet at the top for connection to an air extract system. The glass front is usually vertically sliding but sometimes horizontally sliding. The work surface should have a raised lip around the periphery and a waste outlet for connection to the waste system. Service outlets are mounted on the side walls, with remote controls on the outside of the cabinet, where socket outlets and the cupboard controls are also positioned. The side walls of the cabinet are usually of double-skin construction about 150 mm (6 in) thick, with the supplies to services between the skins and with the skins meeting at the front in a 45° chamfer (for airflow reasons). The socket

Fume cupboards

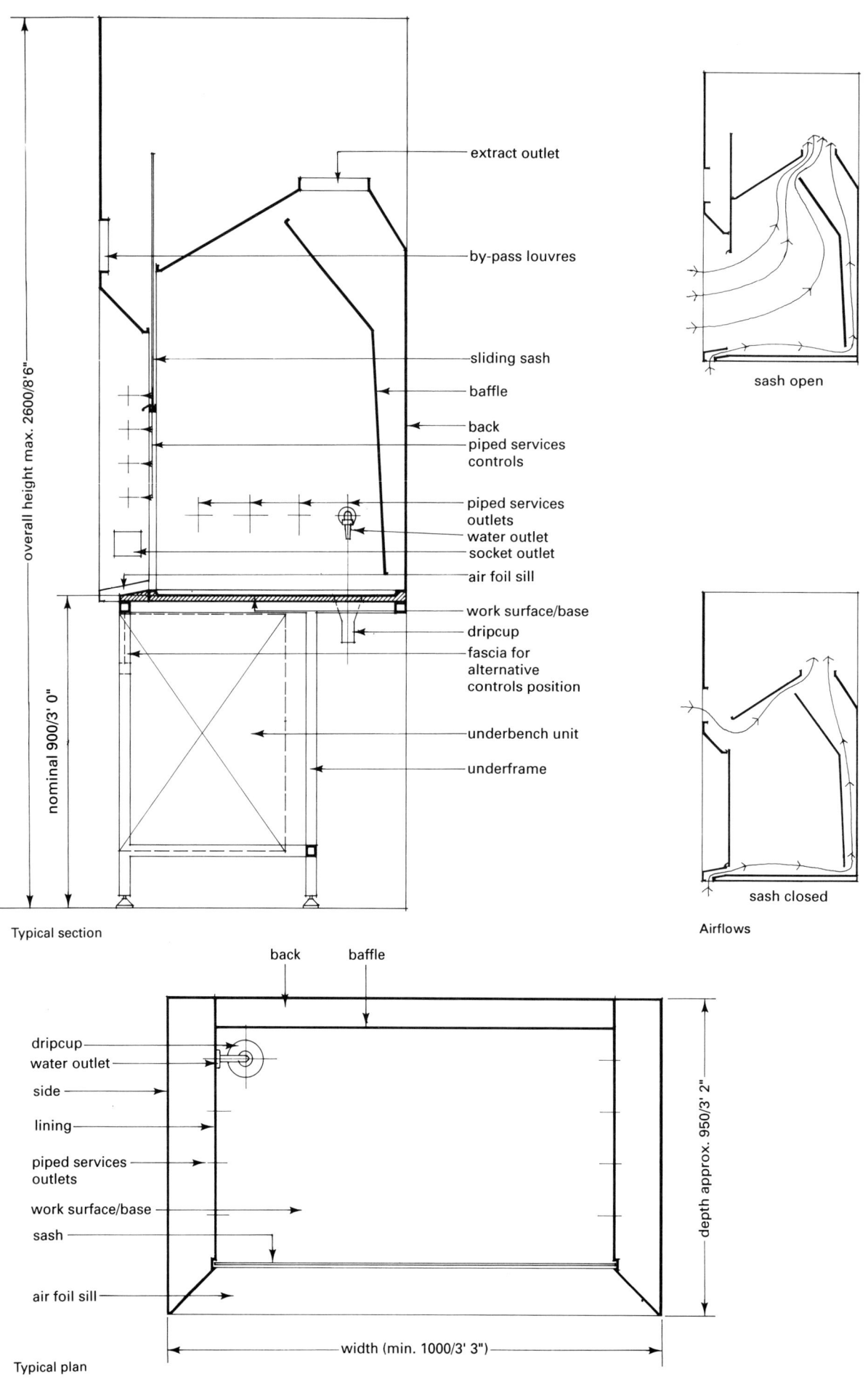

Typical section

Typical plan

Airflows

50 Fitting out laboratory spaces

Fume cupboards

Standard fume cupboard.

Walk-in fume cupboard.

Fume cupboard controls

Fume hood

outlets, remote controls and cupboard controls are usually mounted on the chamfer or on a fascia.

Walk-in or full-height fume cupboards extend to the floor, which usually forms the work surface, and are used where very tall apparatus is being used.

The outer skin does not come into contact with the fumes and is usually of steel or GRP. The inner skin, called the lining, is in contact with the fumes, and the material will depend upon the substances that are to be worked with in the cupboard, as will the material for the work surface. The client should specify the substances to be worked with, and fume cupboard manufacturers will normally propose the appropriate lining and work surface materials to resist these substances. Guidance on these is shown in Technical supplement 11.

The use of perchloric acid in a fume cupboard requires periodic washing-down of the interior of the cupboard and of the ducting system to prevent the formation of dangerous deposits, which can cause explosions.

Fume cupboards are potentially hazardous. They should not be sited near escape routes or high traffic areas nor near doors, and should be a minimum

Fume cupboards and hoods

of 300 mm (12 in) from return walls or columns, for efficient airflows.

Most fume cupboards are about 950 mm (3 ft 2 in) deep and can be accommodated within a floor-to-ceiling height of 2600 mm (8 ft 8 in). The narrowest practical width is 1000 mm (3 ft 3 in), giving a working surface width of 700 mm (2 ft 3 in). Most fume cupboard manufacturers offer widths of 1000, 1200, 1500 and 2000 mm (3 ft 3 in, 4 ft, 5 ft and 6 ft 6 in), and for coordination with bench spaces it is desirable to make fume cupboard widths compatible with the chosen bench module.

Face velocities – velocity of airflow through the front of the cupboard with the sash open, usually at 600 mm (2 ft) opening height – vary with the use for which the cupboard is intended, 0.5 m/s (100 ft/min) being a common face velocity.

The fumes are conveyed in ducts, usually UPVC, to the highest part of the building and discharged vertically through high-velocity outlets to achieve maximum dispersion. The duct from a 1500 mm (5 ft) wide fume cupboard will have a diameter of about 300 mm (12 in), and in multi-storey buildings the extracts can consume a lot of space. The decision on how and where they are to be accommodated can therefore be fundamental to the design and should be taken at design concept stage. If possible, spare space should be allowed for future extracts, in addition to those in the brief, to allow for future lab changes.

Many lab furniture suppliers supply fume cupboards, and some suppliers also undertake the extract fans and ducting, but these are often the responsibility of the main mechanical ventilation subcontractor.

The extract controls must be linked to the supply ventilation system, as the make-up air supply only comes into operation when the fume cupboard is working.

Services spines stop against the sides of fume cupboards, with piped run-outs and wastes continuing through below the work surface, so that the 950 mm (3 ft 2 in) depth will project about 150 mm (6 in) beyond the front of a standard 600 mm (2 ft) deep bench.

Piped services are usually terminated with valves below the fume cupboard for connection by the cupboard supplier to the cupboard outlets. The power supply is similarly terminated in an isolator below the cupboard, for connection to the cupboard by the supplier. An example of a pro forma for listing the fume cupboards on a project together with their requirements is given in Technical supplement 12 and a performance specification in Technical supplement 13. In the UK, brochures on fume cupboards are published by a number of suppliers, a very informative one being that by Morgan and Grundy Limited.

3.4.2 Fume hoods

These are devices mounted above a workplace to remove fumes that may cause a nuisance or discomfort but do not constitute a hazard to health, as in a cooker hood. Because the fumes pose no hazard to health it is usually acceptable to discharge direct to the nearest external air (unlike fume cupboard discharges) through windows on the external wall. Fume hoods may be constructed of steel, aluminium or GRP. The size of the hood will depend upon its function, but over benches it will usually be about 900 m (3 ft) deep and suspended 2000 mm (6 ft 6 in) above floor level, with width to suit the fume producing appliance. Fume hoods can be supplied by the fume cupboard supplier or by the mechanical ventilation subcontractor, as convenient.

Note: In the USA the term 'fume hood' refers to a 'fume cupboard'.

4 Laboratory equipment

4.1 Categories

4.1.1 Category 1: Items supplied by the client and installed by him after hand-over of the building

Items in this category will include bench-standing equipment that will not concern the designer unless it has power requirements in excess of the standard 13 A (in the UK) provision, or other services requirements, such as water, or extract ventilation. Any bench-standing equipment with such requirements must be identified in the brief. This category will include floor-standing equipment such as refrigerators, deep freezers, centrifuges and counters, which concern the designer because of their floor space requirements and which also often have services requirements that call for special provision to be made.

4.1.2 Category 2: Items supplied by client to the contractor for installation by him under the building contract

There is no hard-and-fast rule for the items in this category, as they will depend upon the policies of individual clients. This category is more likely to be used on projects where existing equipment is to be reinstalled, than on new-build schemes.

4.1.3 Category 3: Items supplied and installed under the building contract

Equipment in this category will again depend upon the policies of the client, but items usually included are the glass washer, dryer and autoclave in centralized wash-up rooms.

4.2 Range of equipment

The following list is intended as a guide to the majority of equipment affecting the designer that is likely to occur on a normal laboratory project. It does not pretend to be exhaustive; the suggested space standards and services requirements may vary with the equipment chosen, and should always be confirmed with the client on each project. Width = the front dimension, depth = the front to back dimension. Equipment is floor-standing unless otherwise noted.

4.2.1 Refrigerators

Similar to a domestic fridge, occupying a space 600 mm (2 ft) square on plan and supplied from a 13 A socket. Underbench fridges should be supplied from a socket mounted below the worktop, to avoid trailing cables. (Cat 1)

4.2.2 Deep freezers

The upright type occupy the same space as a fridge and are supplied from a 13 A socket. The −80 °C chest type are about 1700 mm (5 ft 8 in) wide × 750 mm (2 ft 6 in) deep and require a 20 A power supply. (Cat 1)

4.2.3 Centrifuges

Sizes vary, but most can be accommodated in a space 1200 mm (4 ft) wide × 1000 mm (3 ft 3 in) deep and require a 30 A power supply. Some require a cold-water supply and waste connection at near-floor level. (Cat 1)

4.2.4 Scintillation counters

These may be bench- or floor-standing, the latter varying between about 1000 mm (3 ft 3 in) wide × 800 mm (2 ft 8 in) deep and 1500 mm (5 ft) wide × 1000 mm (3 ft 3 in) deep. A 13 A socket supply is usually sufficient. (Cat 1)

Equipment guide

	Equipment	Size Width (mm)	Depth (mm)	Power supply	Services	Category
1	Refrigerator	600	600	13 A	–	1
2	Deep freezer: upright	600	600	13 A	–	1
3	Deep freezer: 80 °C, chest type	1700	750	20 A	–	1
4	Centrifuge	1200	1000	30 A	Some require CW and waste	1
5	Scintillation counter	1000–1500	800–1000	13 A	–	1
6	Incubator	650	650	13 A	CO_2, O_2	1
7	Freeze dryer	1000	600	13 A	–	1
8	Ice-maker	1000	800	13 A	CW, waste	1
9	Dishwasher, domestic	600	600	13 A	CW, HW, waste	1
10	Glass washer	1000	1000	Fused spur	Purified water, CA, drain	2 or 3
11	Dryer	900	800	Fused spur	–	2 or 3
12	Autoclave	800	1200	Fused spur	HW, CA	2 or 3
13	Biological safety cabinet	1500	750	13 A	–	1 or 2
14	Laminar-flow cabinet	1300	750	13 A	–	1
15	Electron microscope	Various		30 A	May require cooling water	1
16	Magnetic resonance imager	Various		30 A	May require CA	1

Laboratory equipment

4.2.5	Incubators	These may be bench- or floor-standing, both usually about 650 mm (2 ft × 2 in) square, requiring a 13 A socket together with a supply of carbon dioxide, and also possibly oxygen. (Cat 1)
4.2.6	Freeze dryers	These are about 1000 mm (3 ft 3 in) wide × 600 mm (2 ft) deep, and require a 13 A supply. (Cat 1)
4.2.7	Drying ovens	These are generally wall-mounted above washing-up sinks; sizes vary. A 13 A supply is required. (Cat 1)
4.2.8	Ice-making machines	These are about 1000 mm (3 ft 3 in) wide × 800 mm (2 ft 8 in) deep, requiring a 13 A supply, cold-water supply and waste connection at near-floor level. (Cat 1)
4.2.9	Stills	These are usually wall-mounted at high level above washing-up sinks, with fused-spur power supply at high level. If a centralized purified water system is installed then it is usual to have the water supply to the still from this system. (Cat 1, 2 or 3)
4.2.10	Dishwashers	A domestic-type dishwasher is sometimes used in labs, 600 mm (2 ft) square on plan, requiring a 13 A supply at low level, hot- and cold-water supplies and a waste connection. (Cat 1)
4.2.11	Glass washer	This is usually in a central wash-up room. Many clients have a preferred make of machine. A suitable size would be about 1000 mm (3 ft 3 in) square on plan and would require a fused-spur power supply, a purified water supply, a compressed-air supply and a drain connection. (Cat 2 or 3)
4.2.12	Dryer	In association with the glass washer, this is about 900 mm (3 ft) wide × 800 mm (2 ft 8 in) deep, requiring a fused-spur power supply. (Cat 2 or 3)
4.2.13	Autoclave	This is in the central wash-up room, about 800 mm (2 ft 8 in) wide × 1200 mm (4 ft) deep, requiring a fused-spur power supply, hot water and compressed air. (Cat 2 or 3)
4.2.14	Microbiological safety cabinets	These are designed to provide protection for both user and the environment from the hazards associated with handling dangerous biological material, and are in three classes (see Technical supplement 4). They may be bench-mounted or supplied with a tubular stand. Classes 1 and 3 require extract ducting to the external air, Class 2 are usually laminar-flow recirculating cabinets, i.e. no extract ducting to the outside. Sizes will vary with manufacturer, but should be in the order of 1500 mm (5 ft) wide × 750 mm (2 ft 6 in) deep × 1500 mm (5 ft) high (height includes duct connection). The cabinets require a 13 A socket supply. Classes 1 and 3 exhaust through HEPA filters and may be discharged adjacent to opening windows. (Generally Cat 1 but may be Cat 2)
4.2.15	Laminar-flow cabinets	These are essentially clean-air cabinets for applications demanding high standards of air cleanliness (e.g. tissue culture areas) in which room air is drawn through filters into the cabinet and may then exit either through the front of the cabinet or through the top. The cabinets may be bench-mounted or may be supported on stands. Sizes will depend upon the manufacturer and the use envisaged, but will be about 1300 mm (4 ft 4 in) wide × 750 mm (2 ft 6 in) deep × 1200 mm (4 ft) high. A 13 A power supply is required. (Cat 1)
4.2.16	Electron microscopes	The sizes and servicing requirements will vary with the machine used and should be specified in the brief. The machines usually have their own transformers, which should be outside the EM room and require a 30 A supply. Cabling from the transformer to an isolator in the EM room is usually installed by the contractor under the building contract, with the machine supplier cabling from the isolator to the machine when installing. Cooling water may be required. (Cat 1)

Lab equipment

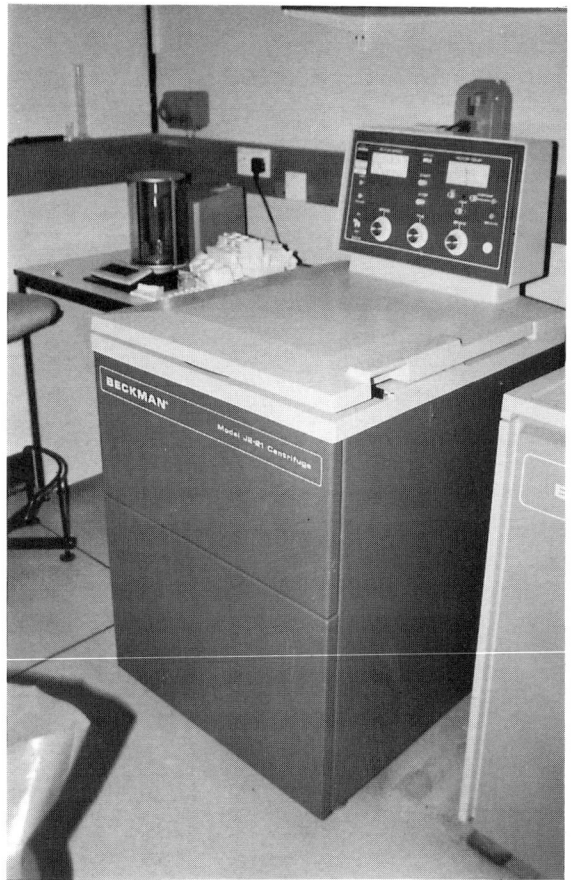

Centrifuge

Bench centrifuges

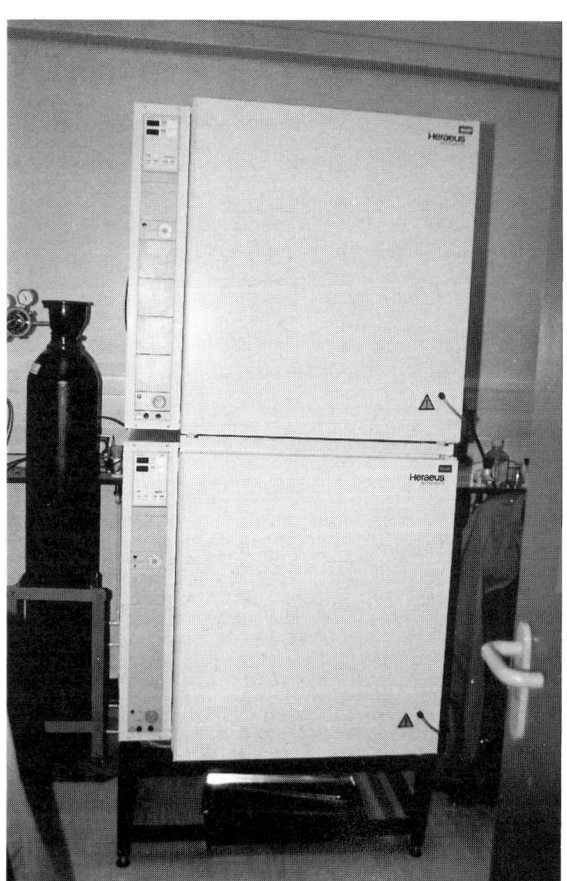

Incubators

Glass washer

Lab equipment

Autoclave

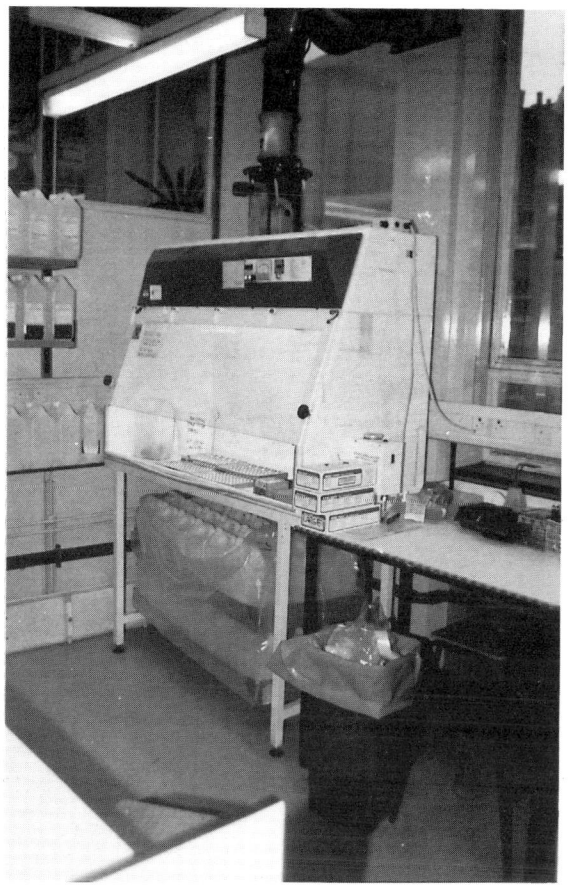

Microbiological cabinet

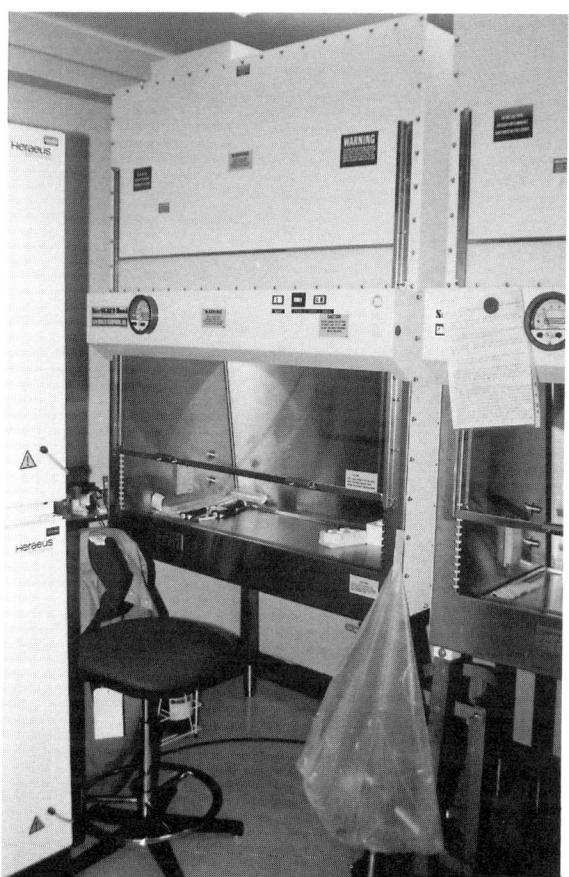

Laminar-flow cabinet

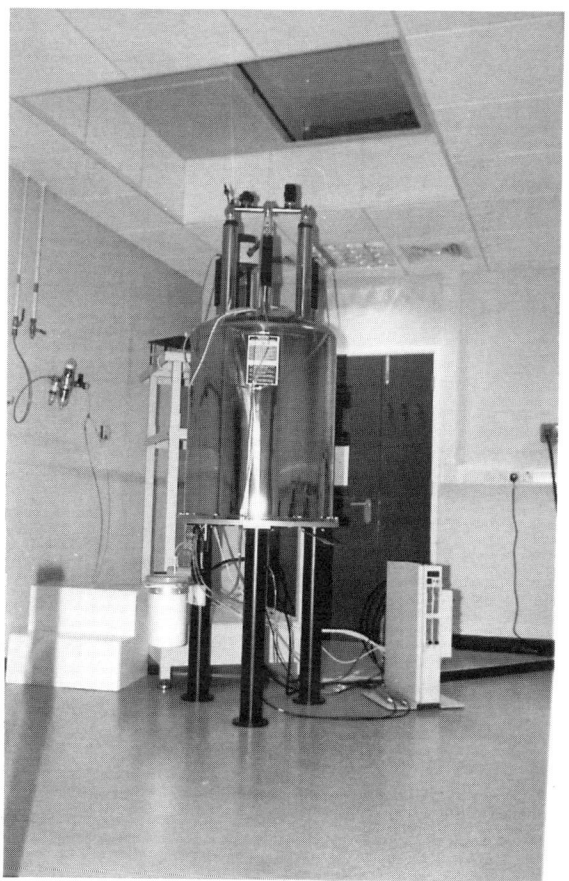

MRI magnet

Range of equipment

4.2.17 Magnetic resonance imagers

Requirements will depend upon the equipment chosen, which usually consists of the magnet itself and the computer console, and may also comprise a permanent magnet. A 30 A power supply is usual and compressed air may be required. (Cat 1)

4.2.18 Animal cages

These are usually supplied by specialist cage-making firms, who produce cages for mice, rats/hamsters/gerbils/guinea pigs, rabbits, cats, dogs and primates. Most clients have a favoured supplier, who will give particulars of cages for the species to be housed. (Cat 1)

The following approximate cage sizes are for guidance only, are for mobile floor-standing units, and should be confirmed with the chosen supplier.

- *Mouse* – mobile rack, 56 cages, 1400 mm (4 ft 8 in) wide × 360 mm (1 ft 2 in) deep × 1500 mm (5 ft) high.

- *Rat/hamster/gerbil* – mobile rack, 24 cages, 1250 mm (4 ft 2 in) wide × 460 mm (1 ft 6 in) deep × 1600 mm (5 ft 3 in) high.

- *Guinea pig* – mobile rack, 24 cages, 1250 mm (4 ft 2 in) wide × 460 mm (1 ft 6 in) deep × 1600 mm (5 ft 3 in) high.

- *Rabbit* – mobile rack, 6 cages, 1420 mm (4 ft 9 in) wide × 510 mm (1 ft 8 in) deep × 1720 mm (5 ft 9 in) high.

- *Cat* – mobile rack, 3 cages, 1140 mm (3 ft 10 in) wide × 590 mm (1 ft 11 in) deep × 1880 mm (6 ft 3 in) high.

- *Dog* – mobile rack, 2 cages, 770 mm (2 ft 7 in) wide × 1070 mm (3 ft 7 in) deep × 1830 mm (6 ft 1 in) high.

- *Primate* – mobile single cage, 780 mm (2 ft 7 in) wide × 1000 mm (3 ft 3 in) deep × 1500 mm (5 ft) high.

4.2.19 Fire extinguishers

The types and positions will be decided by the fire officer. (Cat 1 or 2)

Fire classifications with suitable extinguishers for each are shown in Technical supplement 20.

5 Laboratory services

5.1 Concepts and strategies

This chapter is intended to inform those coming fresh to lab work of the range and nature of engineering services required in labs. It does not provide detailed information on the design of engineering services, which is properly the field of expertise and responsibility of the building services engineer, and applies to the services installations of all building types.

The servicing concept should be formulated and presented to the client for agreement as part of the overall design concept. It should include 'in principle' proposals for mains intake positions, plant room positions, the principal distribution routes, the extent of mechanical ventilation and air conditioning, fume cupboard extract and make-up air systems. It will provide the agreed framework within which each member of the design team will evolve strategies for the particular aspect of the project for which he or she is responsible.

The engineering services in a heavily serviced laboratory building can cost more than the building itself. To ensure that the services installation is cost-effective it is necessary to adopt a servicing concept that will facilitate both the early and easy installation of the services in the first instance, and also provide the accessibility and flexibility to enable the original installation to be maintained easily, and to be added to or modified as new needs arise. There is little point in providing a building fabric that will accommodate change if the services installation inhibits such change.

If maximum flexibility to reallocate lab modules is a requirement then each module must have its own electrical supply and, if labs are to be mechanically ventilated, each module must have its own supply and extract.

To ensure that services within each lab module are sufficiently flexible to effect change the design must allow for:

- the addition of extra bench outlets from the installed piped services;
- the installation of additional piped services, with their outlets;
- the addition of extra power outlets;
- the addition of a fume cupboard.

Additional piped services outlets and power outlets, with their relatively small-bore piping and cables, should not present serious problems. But a design that allowed for the addition in each module of a fume cupboard with its extract duct to the top of the building, together with the supply of the necessary make-up air, could be very expensive. Such total flexibility may not be justified, and it may be sensible to design for a limited number of designated modules where the facility to add fume cupboards would be available. See Technical supplement 16 for a list of services into each lab module that must be allowed for.

Piped services distribution systems

System	Advantages	Disadvantages
Wall-mounted below services	Low installation costs. Good accessibility. Easily modified and modifications only involve lab concerned. No penetration of floor (leaks). Pipes screened by underbench units	Not suitable for island benches
Pipes suspended from ceiling and dropping to individual bench	Low installation cost. Good for serving island benches. Accessible (less so if within ceiling void). Modifications only involve lab concerned	Pipe runs obtrusive unless suspended ceiling used (cost). Drop pipes obtrusive unless boxed (cost)
Pipes on ceiling of floor below, rising through floor slab to feed lab above	Good for serving island benches	Pipe runs obtrusive on floor below (unless suspended ceiling). Holes to be formed through floor slab. Pipes only accessible from floor below. Possible leaks to floor below. Modifications affect floor below
Pipes in floor trench (ground floor only)	Good for serving island benches. Modifications only involve lab concerned	Benches must occur over trenches. Joints between trench covers and floor covering cause problems. Poor accessibility and modifications disruptive

5.2 Mains supplies

5.2.1 Electricity

The normal practice in the UK is for the electricity supplier to require an estimate of the electrical load of the new building, which will be the responsibility of the building services engineer. The supplier will be responsible for supplying and laying the necessary cables into the new building to a meter and an isolator, which he will also supply and install. The building contract will include for the supply and laying of pipe ducts under the building to accept the cables, and for the supply and installation of the main switchboard and for cabling from this to the supplier's isolator.

5.2.2 Water

The normal practice in the UK is for the water supplier to supply and install a meter at an agreed position, together with all piping to that point, and for the piping from the meter into the building and up to the storage tanks to be included in the building contract.

5.2.3 Fuel gas

The normal practice in the UK is for the gas supplier to supply and install a meter and isolator at an agreed position in or on the building, together with all piping to that point. Where the meter is within the building, the building contract will include supplying and laying a pipe duct under the building to enable the gas supplier's pipe to be inserted.

5.3 Bench services

The bench outlets are the lifeblood of the lab: water, fuel gas, special gases and power, without which it cannot function. Although called 'bench outlets', they are now seldom mounted on benches, but on services spines or bridges as described in Chapter 3.

Power outlets are usually provided in trunking mounted separately above the services spine, or above the services outlets in a services bridge, where their installation does not interfere with the installation of the piped services, and where extra socket outlets can easily be added. See section 5.6.1 for detailed information.

5.3.1 Piped services distribution systems

Piped services run-outs supplying the bench outlets are of small-bore piping and are usually run in a horizontal band below the services spine, in the space between spine and floor, with connections up to each outlet. In bridges the run-outs are run within the bridge itself, where the restricted space limits the range of services that can be provided and makes both their installation and their future modification more difficult.

The system of horizontal run-outs is the norm in labs but there is no similar norm for the method of supplying the run-outs from their main circuits, which may be:
- by vertical drops from overhead to each spine/bridge;
- by vertical rises from below to each spine/bridge;
- from vertical ducts at the corridor or window walls into the end of each run;
- from horizontal runs below cill on the window wall into the end of each run supplied from widely spaced vertical drops.

There are pros and cons to each system, and the method to be adopted will depend upon the servicing concepts and strategies agreed upon for the particular project. As a general rule, supply from underfloor should be avoided because it is a potential source of leaks through the floor, and access for modifications is via the room below and therefore disruptive and inflexible.

5.3.2 Piped services outlets (lab taps)

Specialist firms supply the outlets (also referred to as lab taps or valves), which are available in two forms – bench outlets for mounting on horizontal surfaces, wall outlets for mounting on vertical surfaces – and in different types to suit the different liquids, gases and airs. The illustrations show typical single outlets; double, triple and quadruple configurations are also

Piped services distribution

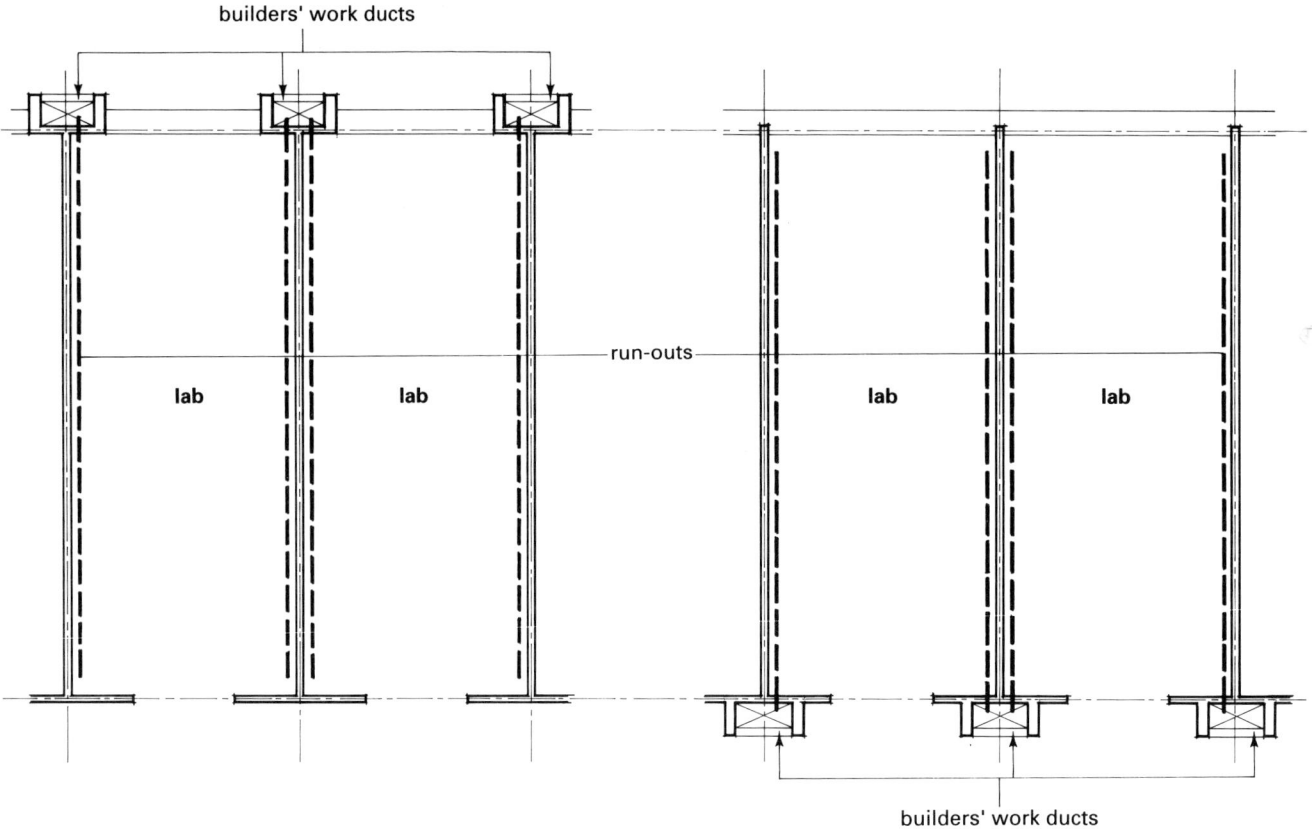

Option 2

Option 1

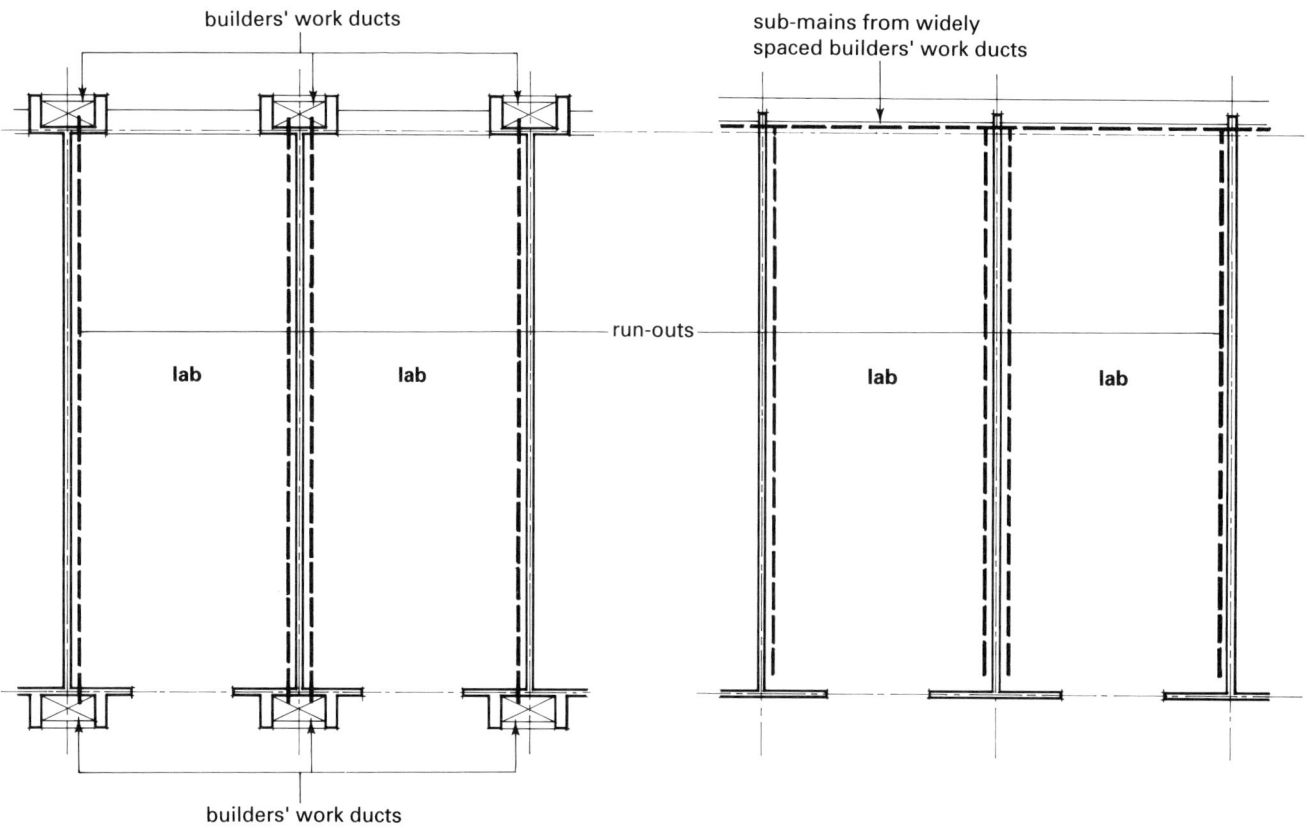

Option 3

Option 4

Laboratory services

Piped services outlets

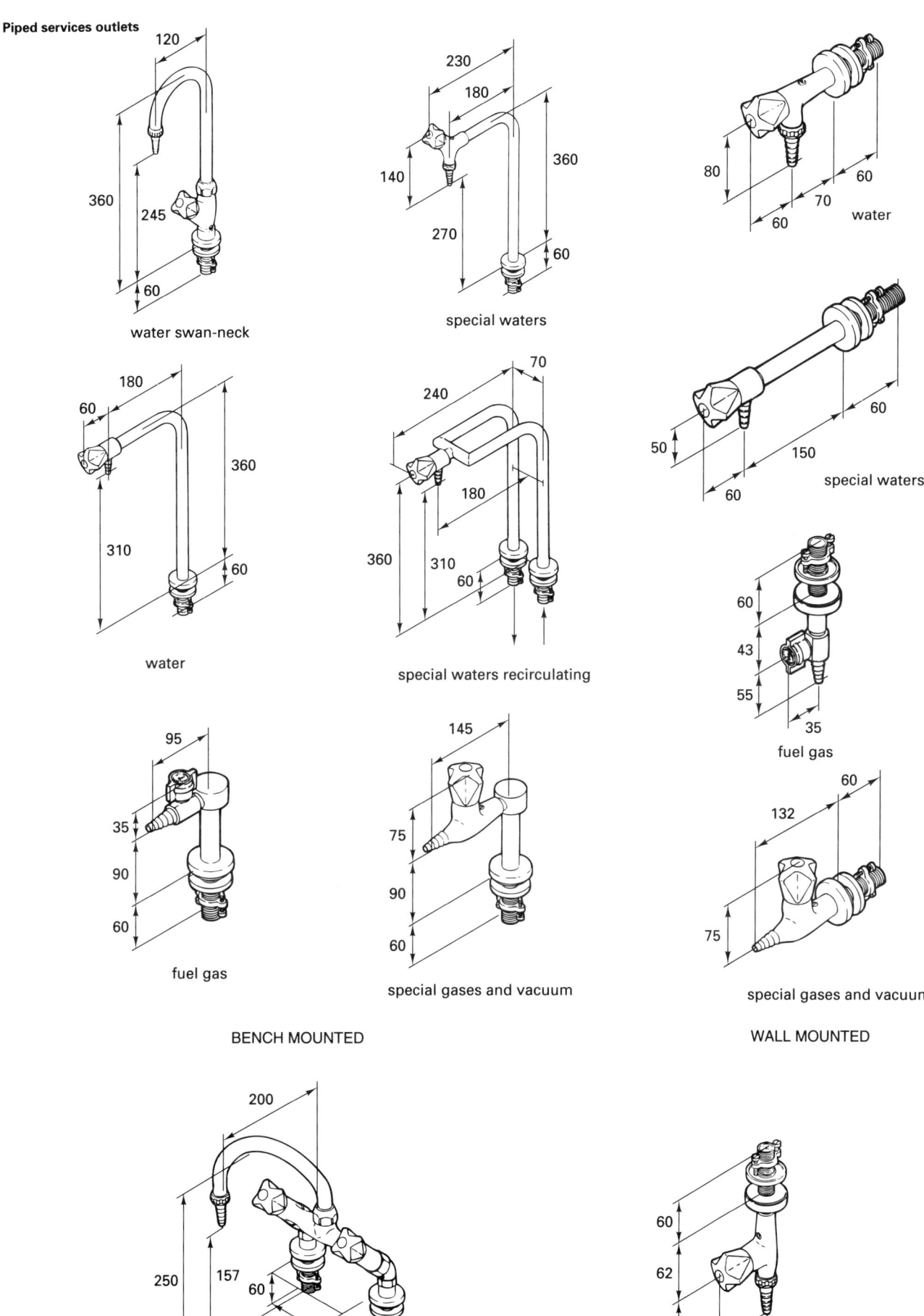

Bench services

obtainable and the client should stipulate which is required. In the European Union the handles are colour-coded in accordance with DIN 12 920 to identify the gas or liquid conveyed: see the chart in Technical supplement 15.

Valves with removable nozzles should be specified, to enable the users to connect piping to the outlets as well as the more usual rubber tubing. Comprehensive ranges are offered by Broen Valves Ltd, Brownall and others.

5.3.3 Piped services types

The following list includes the different types of piped services that are likely to be encountered in lab work. The abbreviations shown are those generally used in floor data plans.

Cold water (CW)

The down-service supply from the storage tanks in the plant room will normally be required to sinks, handbasins and water standards.

Pressure water (PW)

May be required if the down-service pressure is low. Supplied from pumps in the plant room; the required pressure should be stated in the brief.

Drinking water (DW)

Direct from the mains; will be required to specific points such as kitchen sinks and may be required elsewhere, such as drinking fountains and animal rooms.

Hot water (HW)

Will be required to washing-up sinks and handbasins.

Purified water

The three processes used to purify raw water are reverse osmosis, deionization and distillation.
- *Reverse osmosis (RO)* separates the water from impurities and removes 99% of bacteria, viruses and organic substances and 95% of dissolved solids.
- *Deionization (DEI)* uses resins to remove all dissolved ionizable substances from water.
- *Distillation (DIS)* is a double phase change from liquid to vapour to liquid, removes dissolved mineral salts and produces sterile water.

RO water is acceptable from a central system, but the contamination dangers inherent in long pipe runs and remoteness of control mean that the other two are not. A system often used is to pipe RO water from a central plant to the point of use, generally at a sink, where it may be used for rinsing glassware or may be further purified for use in lab procedures, either by proprietary cartridge system for deionized water or by local still for distilled water, both mounted on the wall above the sink. Outlets for purified water are called 'special water' outlets and are either in stainless steel or lined with plastic tube. Recirculating outlets are available to avoid deadlegs in pipe runs.

Fuel (town) gas (G)

Required in some labs, but not to the extent previously seen.

Compressed air (CA)

Often required in chemical and biological labs (e.g. for driving rotary instruments); may be piped from central plant or by local electrically operated pump on the bench. The pressure required should be stated in the brief.

Vacuum, or suction (V)

Central plant can give rise to problems regarding disposal of the effluent and it is often generated locally on the bench by water-operated or electrically operated pump, when the effluent is discharged down a dripcup.

Laboratory gases

A wide range of special gases is available, including nitrogen (N), nitrous oxide (N_2O), carbon dioxide (CO_2), carbogen (carbon dioxide/oxygen mixture) (CO_2/O_2), and helium (He). Requirements should be defined at briefing and a decision subsequently made on which will be piped in from a central cylinder store and which will be supplied from a cylinder in the lab at the point of use.

All special gases use the same pattern of outlet except oxygen, which requires joints to be silver-soldered and the valve degreased.

Acetylene (ACE)

This is sometimes required and is subject to special restrictions: piping must be in steel, exposed to view, and there are requirements for isolating valves, non-return valves and flame arresters. The installation has to be examined and certified at least once every 12 months. There are also restrictions on the storage of acetylene cylinders. The installation is subject to the Compressed Acetylene Order 1947 and there is a 20 Point Guidance Code. Approval is required from HM Explosives Inspectorate, Health and Safety Executive. Copper alloys containing more than 70% copper should not be used in any part of outlets except burning nozzles.

Steam (S)

This is not often required; if it is called for and there is no steam supply available, then it is usually supplied from a steam generator in the plant room. Bench outlets for steam are of a special diaphragm or needle type.

Pressure gas control consoles

These are sometimes required in labs, to enable the users to set the precise pressures required at the gas outlets, for wall or services spine mounting. A suitable type is supplied in the UK by Mason Nordia Ltd.

Pipe materials

The majority of piped services are usually conveyed in either copper or plastic piping, services with specific requirements being:
- *purified water* – must be in plastic or stainless steel;
- *acetylene* – must be in steel;
- *steam* – must be in steel.

5.4 Dripcups and wastes

5.4.1 Dripcups

Dripcups (or tundishes) are waste outlets on services spines into which fluids can be tipped and equipment can be drained, and are usually used in conjunction with a swan-neck water standard discharging over them. They may be circular or oval, and are generally recessed flush into a rebated hole in the services spine, bedded in silicone sealant, and require a plywood support box under the top against which the backnut can be tightened. Suitable dripcups are Vulcathene's 501 Small Circular Drip Cup and 497 Small Oval Drip Cup, in polypropylene.

5.4.2 Wastes

Wastes are required from sinks, dripcups, fume cupboards, etc. They are usually 50 mm in diameter and run below the bench services run-outs under services spines. They may be in polypropylene (as Vulcathene 'mechanical'), polyethylene (as Vulcathene 'fusion'), or borosilicate glass. Polypropylene is resistant to most fluids used in labs; its mechanical joints enable easy modification and it only requires supports at 1220 mm (4 ft) centres. Polyethylene is slightly less resistant, has welded joints and requires continuous support but is favoured for radioactive wastes. Borosilicate glass is resistant to virtually everything, has mechanical joints and only requires supports at 1220 mm (4 ft) centres, but is more expensive, can be susceptible to mechanical damage and is normally only installed by specialist suppliers. The Vulcathene brochure contains chemical resistance charts showing the effect upon its products of a wide range of chemicals (see Technical supplement 18).

There are three main categories of waste: normal, radioactive and biohazardous.

It is usual for a run of normal lab waste pipe to discharge into dilution recovery traps or catchpots, before connection to floor outlets or stacks. A suitable catchpot is Vulcathene 910V or 910G (the latter is in borosilicate glass for identification and collection of valuable solids).

Dripcup with equipment drains

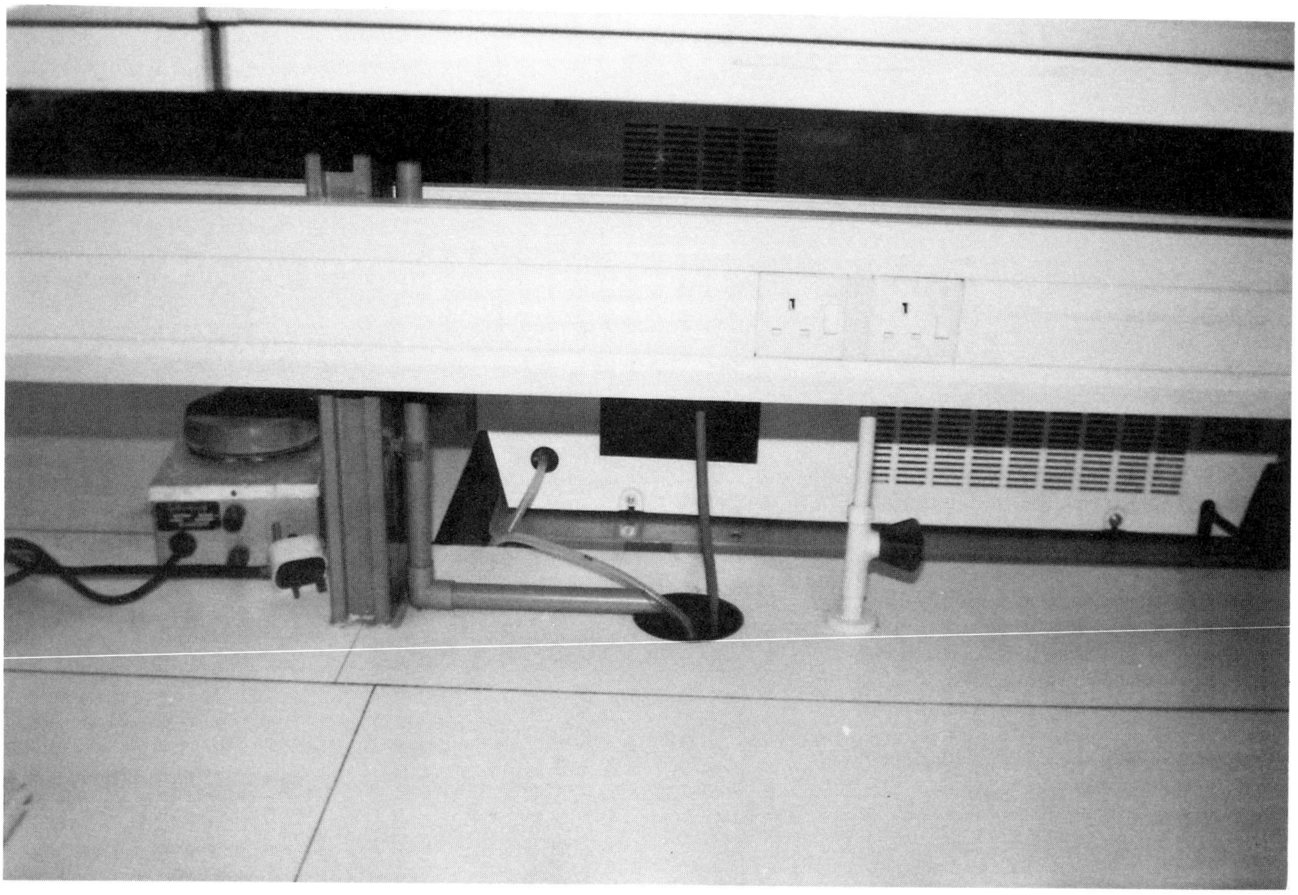

Pure-water unit over sink unit

Floor outlet

Laboratory services

Dripcups and dilution recovery traps

Small circular dripcup

Small oval dripcup

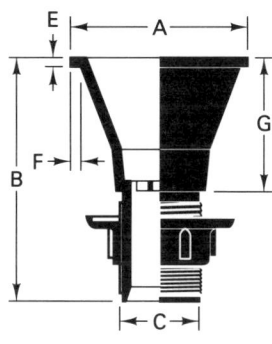
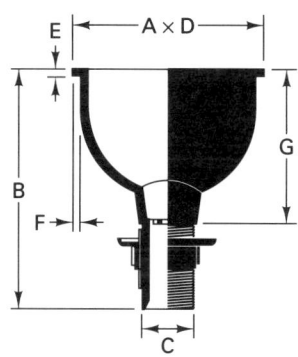

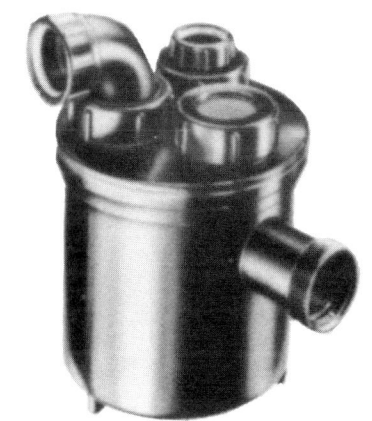

cat. No 501	cat. No 497
A 102	A 175
B 136	B 216
C 1½″ BSP	C 1½″ BSP
E 5	D 102
F 6	E 6
G 76	F 6
	G 143

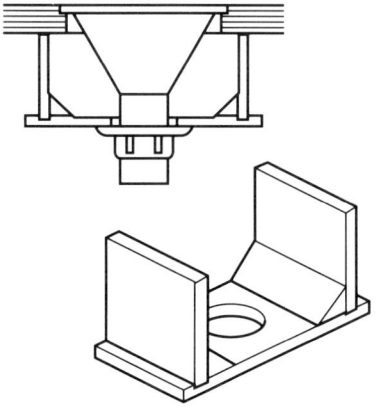

Plywood box for securing dripcups

cat. No 501
A 194
B 235
C 264
D 152

Dripcups

Catchpots

Dripcups and wastes 67

Radioactive wastes should be taken direct to the stack/floor outlet without receiving wastes from other fittings *en route*. If the waste is strongly radioactive then the local authority should be consulted.

Biohazardous waste requirements should be discussed with the local authority.

5.5 Drains

A waste connection point should be provided on each long side of each lab module. On upper floors this can be by means of a waste stack and on ground floors by a floor outlet, with solid cover with a threaded 50 mm (2 in) connection to take the waste pipe (a suitable fitting being the Wade W1200 series Floor Drain). Floor drains will be required in animal rooms and may be called for in other rooms such as wash-ups (a suitable drain is the Wade W1200 series Floor Drain with open grille cover).

Normal lab wastes should occasion nothing out of the ordinary in drain requirements, but the authority responsible for receiving the effluent (usually the water authority) will be interested in what is to be discharged and may wish to take samples of the effluent from the labs from time to time for testing. The authority may require dilution chambers for radioactive drainage and similar provision for biohazardous drainage.

Most items of plant have waste outlets, and in plant rooms floor drains are required at regular intervals, into which the wastes can be discharged.

5.6 Power, lighting and communications

5.6.1 Power

Supply

For maximum flexibility it is necessary for each lab module to be supplied from its own consumer unit, so that modifications can be carried out to one module without affecting the supply to other modules.

Small power

All labs make use of large numbers of bench-mounted electrical appliances, generally of low load, and so require a plentiful supply of socket outlets to avoid the use of double adaptors and trailing cables. These outlets are usually provided by means of three-compartment dado trunking mounted above the services spines, in either PVC or steel, the main compartment for power and the others for telephones and data. The busbar type of trunking provides maximum flexibility in that it enables extra socket outlets to be provided without additional wiring merely by plugging a socket outlet in to contact the busbars (a suitable type being MK Powerlink). A useful rule of thumb for small power in labs is to provide one twin 13 A socket outlet per metre run. Turret or pedestal-type fittings are available where socket outlets are required on benchtops, in either single- or double-sided versions. Power is sometimes required at high level where there are no support posts or benching, to supply floor-standing equipment. The power trunking is then fixed at ceiling level or, in labs with a very high ceiling height, suspended at doorhead level. An example of the latter is illustrated, in a hospital chemical pathology lab. Here it is used to supply services to analysing machines, and in addition to socket outlets and 30 A isolators it also carries cold and RO water, together with refrigerant piping and drains from ceiling mounted cooling units.

Non-standard power

Three phase power, low-voltage power, 20 A or 30 A power for particular items of equipment may be required and should be identified in the brief. The first two will require cabling separate from the small power trunking, but the other two can usually be supplied from the trunking, with isolators of the

Small power

Consumer units

Busbar power trunking

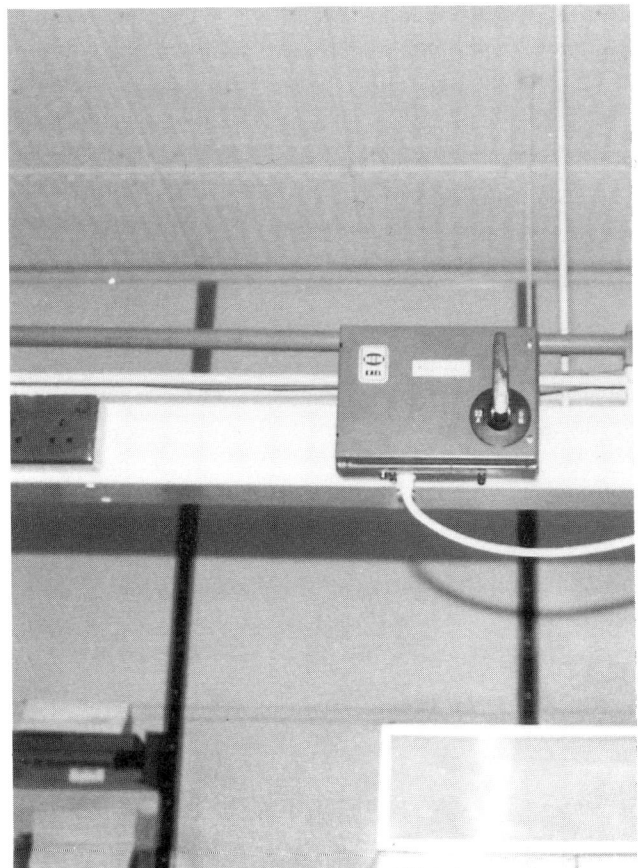
High-level trunking

30 A isolator

Power lighting and communications

appropriate size mounted on conduit 100 mm (4 in) above the trunking (the 100 mm gap is to enable the equipment to be connected to the underside of the isolator).

Dedicated power supplies, serving only a particular piece of equipment or room, are sometimes required and should be identified in the brief, as should stabilized supplies, which are also required occasionally.

Earths

Dedicated earths, serving a particular bench run, are sometimes required and should be identified in the brief. They will require a 50 mm (2 in) PVC pipe duct with easy bend under the ground-floor slab to about 300 mm (1 ft) beyond the external wall, to provide entry into the building for the cable from the earth rods.

5.6.2 Lighting

Labs are customarily lit by means of fluorescent luminaires, parallel with the run of the bench that they serve and about 900 mm (3 ft) from the wall. A level of illumination at bench level of 500 lux is usual but the specific requirement should be stated in the brief. Each lighting run should be separately switched. Researchers will generally provide their own local task-lighting from the small power outlets.

Special lighting may be required in addition to the general lighting in some areas, such as ultraviolet in tissue culture labs and safelights in darkrooms, and these should be identified in the brief. Emergency lighting will be required to light escape routes. Direct penetration of sunlight into labs should be screened to prevent glare.

5.6.3 Communications

Telephones

Electronic exchanges occupy little more space than a typewriter, so that space requirements are not normally a problem. Telephone points should be identified in the brief, and the client will decide the system to be used and make arrangements with the telephone company regarding the equipment to be supplied and the installation of that equipment. Carcassing for the installation is, in the UK, usually included in the building contract in the form of conduit and trunking, with the wiring and the installation of the equipment and instruments carried out by the telephone company under separate contract with the client, but during the building contract.

Data and computer systems

In the UK these are handled in the same manner as telephones. The data compartment of the standard three-compartment power trunking described in section 5.6.1 may not be sufficient to handle the cabling requirements on some projects. Some trunking systems (such as Powerlink) offer an extension trunking, which fits onto the bottom of the standard trunking to provide an additional large cableway for such situations.

Clocks

Clock positions should be identified in the brief, as should the system to be used:

- *impulse/slave* – for use with a master clock (cabling from master to each slave position);
- *synchronous* – for direct connection to the mains (power supply to each clock position);
- *battery* – individually battery-powered (no services provision).

5.7 Safety, security and safeguarding of work and specimens

5.7.1 Safety

The two main hazards are from chemicals used in lab procedures and from fire.

Safety
Drenchshower

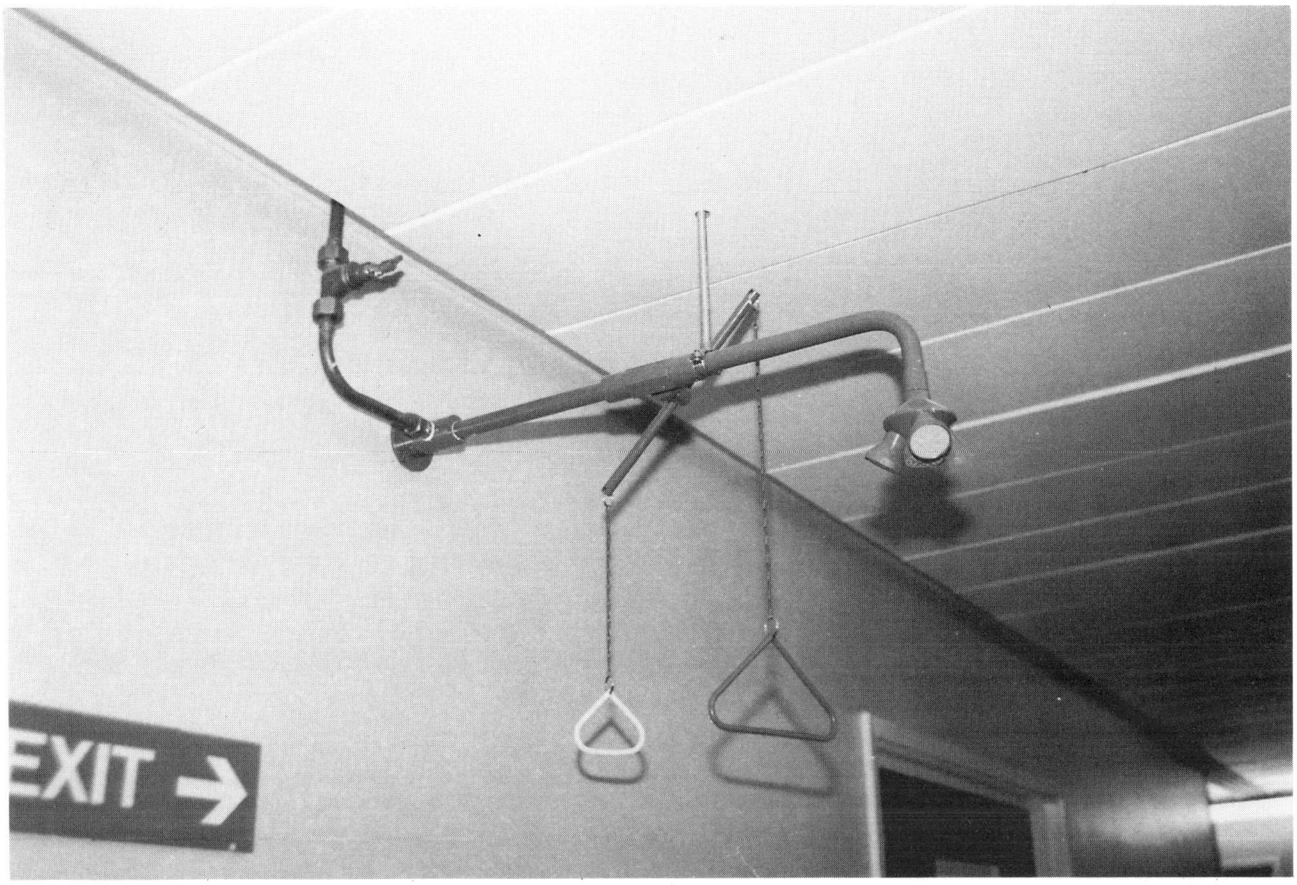

Eyewash

Chemicals

The requirement is to dilute the chemicals by deluging with water. For this, drenchshowers are provided for deluging the body and eyewashes for the eyes.

Drenchshowers are fixed at doorhead height, project about 700 mm (2 ft 4 in), discharge directly onto the floor and are actuated by means of a chain pull (or sometimes by foot plate). Some clients require floor drains to be provided under drenchshowers but, as the latter are only used in emergencies and therefore very infrequently, others consider drains to be unnecessary and prefer to mop up the discharges as they occur (the traps of floor drains may also dry out owing to the infrequent use of the showers). Drenchshowers may be positioned in a lab module but are more usually sited in corridors where one shower can serve a number of modules.

Eyewashes may be placed below drenchshowers or mounted separately.

In the UK a comprehensive range of emergency shower systems is supplied by Broen Valves and another by Brownall.

In chemistry labs it is sometimes a discipline to ensure that nobody should enter the lab unless wearing protective spectacles, which are held in a container on the corridor wall outside the entrance door, to be donned on entering the lab and replaced in the container on leaving.

Fire

The fire alarm systems that are now required by fire officers in the UK comprise automatic fire detectors (heat or smoke) in each lab module and in ancillary rooms, connected to a central alarm system that includes both sounders and lights, and possibly a direct telephone call line to the local fire brigade. The fire officer will mark up a set of drawings showing the locations for sounders and lights. It is usual for the fire alarm system installation to be included in the building contract. The fire officer may also mark up positions for fire extinguishers, or may only decide on these at an inspection of the building when it is almost complete.

Fire protection systems, in which inert gases such as halon discharge when triggered by the fire alarm system, were previously sometimes used to prevent fire damage to expensive equipment in specific rooms, but these have been discontinued in the UK for hazard and environmental reasons.

5.7.2 Security

The activities of animal rights groups have led to the need for a much greater emphasis on security in labs than was previously the case and it is now usual for access to research lab buildings themselves, together with areas within the buildings, to be controlled by means of card-operated lock systems. Openings in the external envelope such as windows and doors may be tripped and connected to an intruder alarm system, together with intruder alarm beams and movement alarms in corridors and rooms.

Research laboratory buildings are often divided internally into security zones, in ascending order of security, with entrance to each zone safeguarded by a door with a card-operated lock, accessible only to those whose personal card has the authority to pass that lock. The locks have a push-actuated release on the inside for exit from the zone. A typical lab building could contain the following zones:

- *zone 1* – general access to the building via the main entrance to areas containing reception, entrance hall, main administrative offices, toilets, and possibly common room, seminar room, stair and lift. Access for visitors is controlled from the reception desk. Also in zone 1, service access into the building via the service entrance into the service lobby;
- *zone 2* – general laboratory areas, accessed from zone 1 into the lab corridor(s);

Communications

Extension trunking

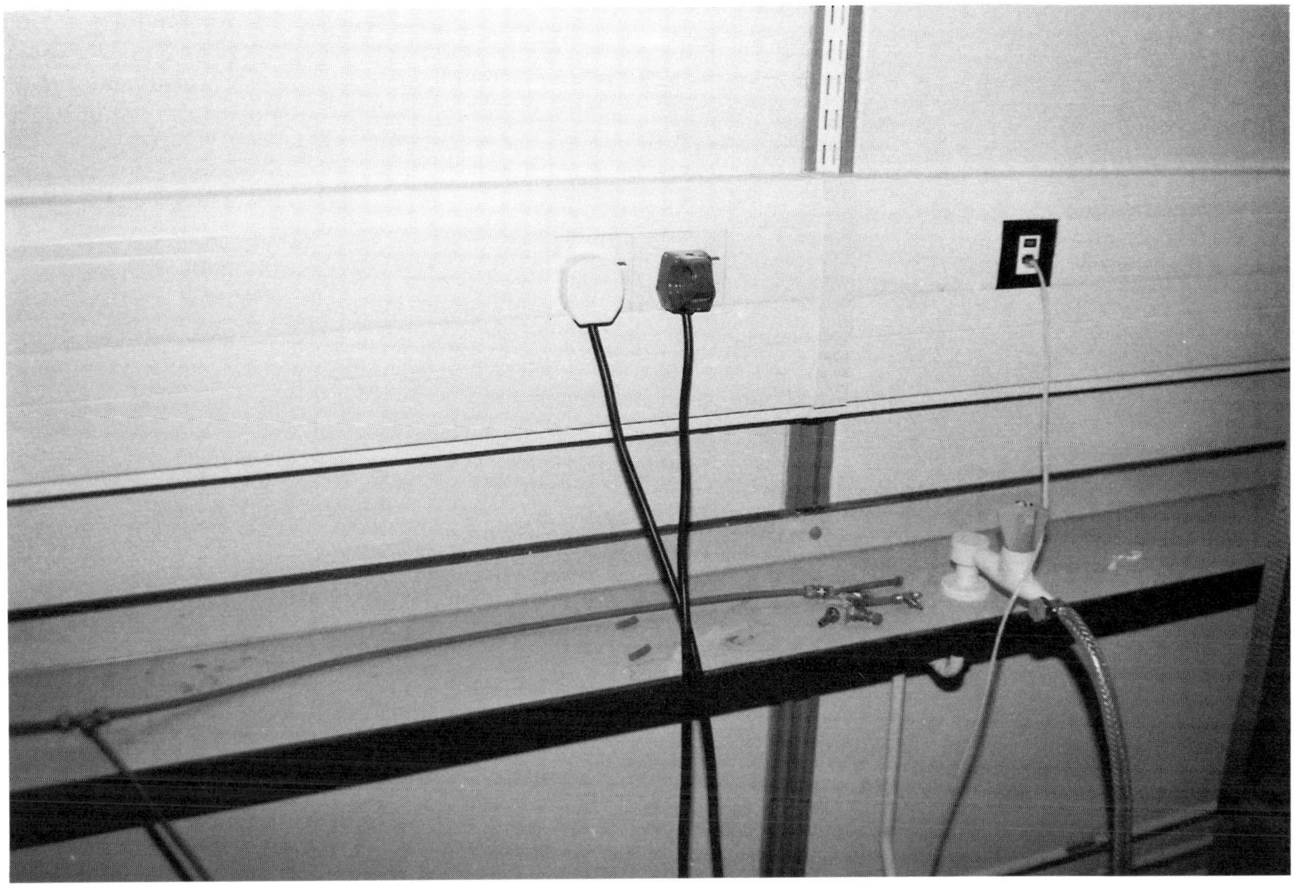

Safety and security

Alarms panel

Safety and security of work and specimens

- *zones 3, 4, et seq* – specific areas that are not accessible to all lab staff but are restricted to specifically designated staff, such as animal areas, biohazardous areas and radioactive areas, accessed from zone 2.

The entrance hall will usually contain a reception desk monitored either by a staff member or by security firm staff, which will control access to the building and where the telephone exchange will be situated. If CCTV cameras are installed then their monitor screens will be at the reception desk, and if there is a main alarms panel (containing fire alarms, alarms from the security system and from the equipment alarms systems) then it will be located at the reception desk position.

Security systems are available in varying degrees of sophistication from specialist security firms, who will advise, supply and install the system, usually under direct order from the client outside the building contract. Carcassing in the shape of conduit or trunking is required for cables to each card lock, trip and component of the system; in the UK this is usually supplied and installed under the building contract.

CCTV cameras may be installed to scan the outside of the building and security firms are sometimes engaged to patrol the inside of the building at night and to monitor the TV screens.

The degree of security required should be discussed and decided on at briefing.

5.7.3 Safeguarding work and specimens

Equipment alarm systems are often required to indicate the malfunctioning of equipment that could result in the loss or destruction of valuable work and specimens, as in cold stores for example. A stand-by electricity supply facility is often required to prevent the loss of specimens or work in progress in the event of a mains failure by supplying designated socket outlets, the environmental services to particular rooms, etc. It involves cabling from the designated points to a switchboard adjacent to the stand-by plant site. A stand-by electrical plant may be provided, which switches on automatically, or a plant may be hired when notice of an intended power cut is given.

5.8 Environmental services

5.8.1 Ventilation

In the UK and similar temperate climates, openable windows, combined with central heating and high standards of thermal insulation, will often maintain satisfactory environmental conditions, especially in office areas. See the external design temperatures table in Technical supplement 22 for comparative temperatures throughout the world.

The decision on whether natural ventilation will be satisfactory in labs, or whether mechanical ventilation is required, will depend upon the purpose for which the lab is to be used. The client will know the purpose and should therefore be in a position to make a decision at briefing. The decision should be taken before any design work is undertaken because it can have a fundamental effect upon the building form, both in plant room space requirements, duct space and, crucially, in making deep-plan solutions feasible. It is likely that some rooms will need mechanical ventilation in a project where natural ventilation is acceptable for most lab spaces, but this should be obvious from the brief for these rooms.

The general mechanical ventilation/air-conditioning systems for laboratory buildings do not differ significantly from those of other building types with the exception of fume cupboards, which are unique to lab buildings.

It is usual for each fume cupboard to be connected to its own separate duct and fan with the discharge vertically into the air above roof level. Horizontal duct runs between cupboard and vertical duct should be as short as

Ventilation

Make-up air-handling unit in naturally ventilated lab

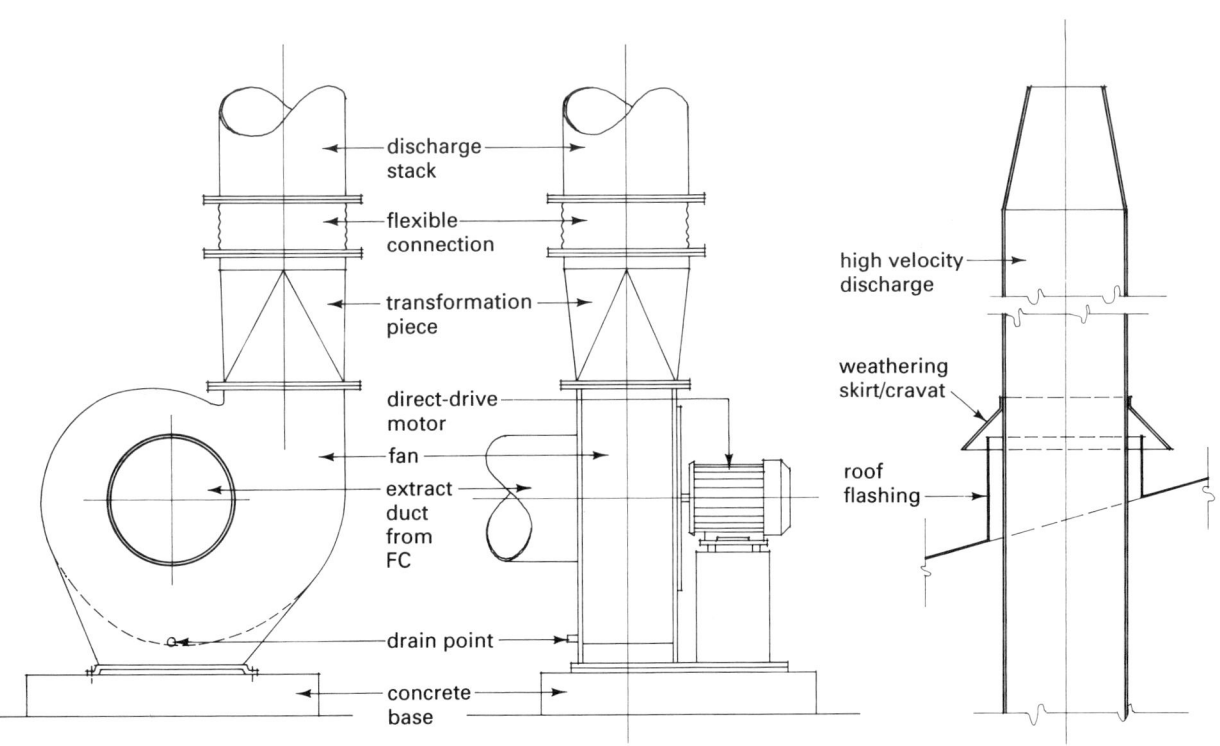

Fume cupboard extract fan

Fume cupboard high-velocity discharge

Environmental services

practicable, so vertical duct space is required close to any lab in which a fume cupboard occurs. The extract duct for a 150 mm (5 ft) fume cupboard will be approximately 300 mm (12 in) in diameter, and in a multi-storey building with many fume cupboards the requirement for vertical duct space presented by these can have a fundamental effect upon the design of the building. This is why an 'in principle' decision should be taken at design concept stage on where and how the necessary vertical duct space is to be provided. The extract fans are normally mounted at roof level, exposed if on a flat roof or in a roof plant room, but may also form the junction between horizontal and vertical sections of the extract duct.

In chemistry labs with banks of fume cupboards these are sometimes connected to a common extract duct and fan with balanced extract; that is, with a regulated amount of outside air introduced into the duct to compensate for cupboards that are not in use. This system requires sophisticated controls; the fan must be run continuously, and a fan breakdown (or maintenance) renders all the fume cupboards inoperable.

Rooms containing fume cupboards require a make-up air supply to replenish the air exhausted through the fume cupboard, and in mechanically ventilated rooms this will be through the general room supply. In rooms with a large number of fume cupboards relative to the floor area it will be necessary to supply make-up air from a plenum through a perforated ceiling, to avoid draughts. In naturally ventilated rooms it is usual to use a local air-handling unit mounted at high level within the room to provide make-up air.

5.8.2 Heating

Radiators or convectors can be difficult to accommodate in labs because all available wall space tends to be occupied by fittings and equipment. In naturally ventilated spaces where it is impracticable to use radiators or convectors, heating may be provided by means of fan-coil heaters mounted at high level. In mechanically ventilated labs heating will normally be by means of the supply air, although even in these additional heating may be required in one of the forms mentioned above to overcome local heat losses through the building fabric.

5.8.3 Cooling

Equipment with a high power consumption has a corresponding high heat output. Concentrations of such equipment often require cooling to maintain acceptable temperatures, as will rooms containing a large number of people, such as seminar and lecture rooms. Cooling may be specified in the brief or may arise from the temperature requirements stated.

In naturally ventilated areas cooling can be provided by means of fan-coil units mounted at high level, cooling the room air by blowing it across coils supplied with chilled water from a central plant, or from its own condenser unit positioned outside the building in the nearest available space, to which it is connected by insulated refrigerant pipes. Cooling units have drains which are often taken to discharge into dripcups.

In mechanically ventilated areas cooling will be by coils in the air-supply system, normally supplied with chilled water from a central plant.

The central plant may consist either of an air-cooled condenser direct to atmosphere (a finned coil with a fan to induce airflow, like a car radiator), or a water-cooled condenser in which the refrigerant is cooled by water from a cooling tower.

The water-cooled condenser will usually consume less power and give more efficient refrigeration, but is more complicated and the cooling water circulation is an open system, with the drawbacks of dirt, corrosion and water treatment and, in recent years in the UK, legionnaires' disease. The bigger the cooling requirement the higher will be the power savings, and the drier the climate the higher will be the efficiency of the cooling tower. The cooling tower(s) must be outside the building envelope exposed to the external air, but the condenser can be within the plant room. The air-cooled condenser is a simple closed-circuit system that lends itself to packaging and is very suitable for UK climatic conditions. It must be outside the building envelope, exposed to the external air. A central system may require a number of condensers, depending upon the cooling load.

Local condensers serving individual cooling units are normally of the air-cooled type and may be mounted on flat roofs or on the vertical face of the building.

Cooling is expensive, and it is usual to restrict it solely to those areas where it is a functional necessity.

5.8.4 Air conditioning

Mechanical ventilation with heating, cooling and humidity control is normally restricted for cost reasons to areas in which it is a functional necessity. Animal rooms are almost always air-conditioned; other areas that require air conditioning should be identified in the brief.

5.8.5 Controls

The many different services and items of plant encountered in lab buildings require a corresponding multiplicity of controls. Water pressures, upper and lower temperature limits, airflow rates, relative humidity percentages, etc., all have to be monitored and controlled. Controls consist essentially of sensors, manometers, thermostats and valves, which have to be fitted into or onto already installed pipework, ductwork or cabling.

Control systems are usually complicated. They can take a considerable time to install and commission because they are fitted to existing installations, often require both mechanical and electrical work and also involve an additional specialist. An adequate time allowance is not always made in programmes to cover either the installation or the commissioning.

In the UK, controls are usually supplied under the building contract by a specialist controls company and installed under its direction by the mechanical or electrical subcontractors, under the building contract.

Partition system

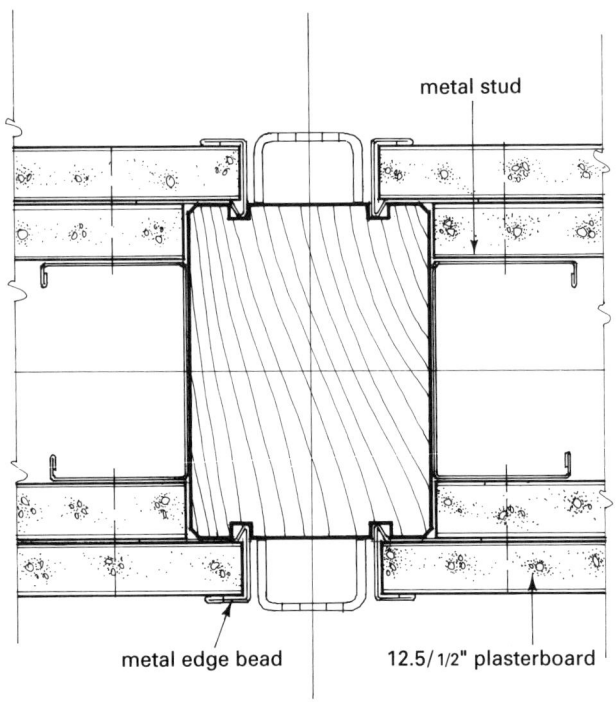

Solid post

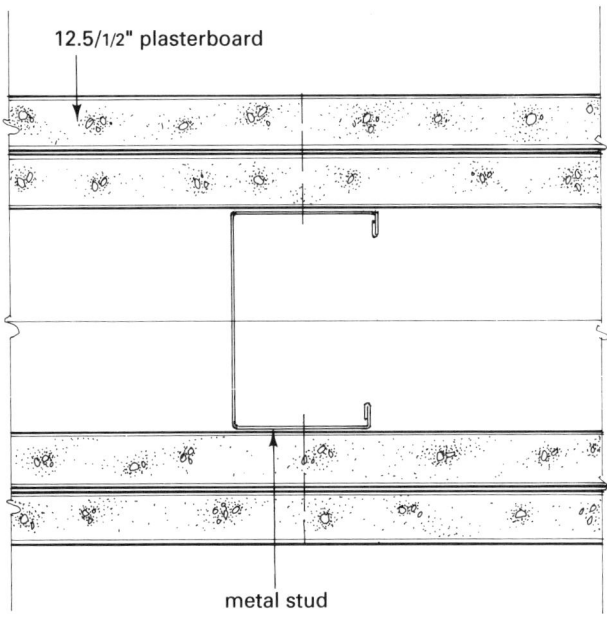

Solid, intermediate stud

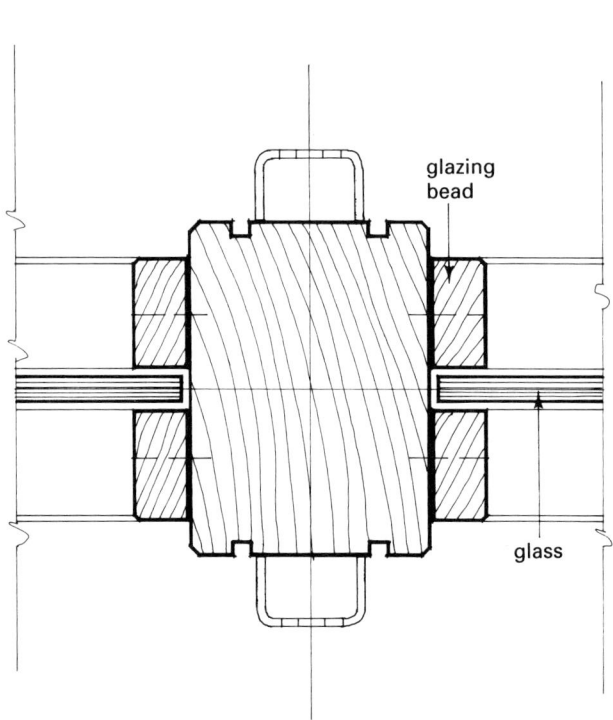

Glazed post

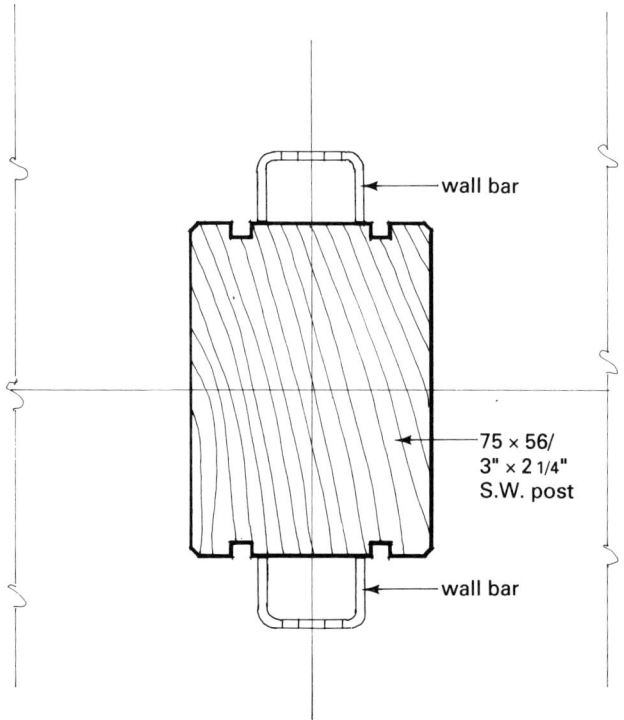

Open post

6 The building fabric

6.1 General

The structure and external fabric of laboratory buildings need not necessarily differ from those of other comparable building types. It is in the internal partitions, finishings and fittings that there are specific requirements for labs. The finishes for rooms other than those particular to this building type will be as is usual for such rooms and have not been covered here. All finishes should of course be approved by the client.

6.2 Partitions

6.2.1 Systems

Partitions for laboratories differ from those of most other building types in one important aspect: more things are hung from them and they therefore fulfil a much greater support role. When this role is matched with the modular nature of many of the items that they support, together with the need to alter, modify or remove partitions completely with the minimum of disruption to adjoining areas, then it is apparent that neither traditional block/brick partitions nor metal stud partitions are ideal. A partition system is required that provides support from floor to ceiling at lab furniture module centres and which can consist solely of a line of support posts with nothing between them, or a line of support posts with glass panels between them, or a line of support posts with solid panels between them.

A system that meets these requirements is the example illustrated, consisting of timber posts fixed to floor and structural ceiling at lab furniture module centres, able to be exposed free-standing or to have the spaces between them filled in with glass or solid panels and, where especially clean conditions are required, to be completely covered by the solid panels.

The system shown is a development of one used by the author in a number of lab buildings and shown in the photographs. It consists of timber posts to which the channel and wall bar supports described in Chapter 3, section 3.1, are fixed. It uses plasterboard for the solid panels. The studs to support the plasterboard panels may be of the proprietary metal channel type as shown, or may be in timber. The system may use single- or double-layer plasterboard. The plasterboard is used drywall for direct decoration in normal laboratory conditions and plastered with a thin finishing plaster where cleaner room conditions are required.

The system has a ½ hour fire resistance with single-layer construction and 1 hour with double-layer. The sound insulation properties are: single-layer construction 38 dB (44 dB with 25 mm (1 in) glass-fibre mat in the cavity); double-layer construction 45 dB (51 dB with glass fibre mat).

6.2.2 Finishes

For most standard labs a plastered or similar surface such as plasterboard finished with vinyl emulsion will be perfectly adequate; the walls are usually

Partition system

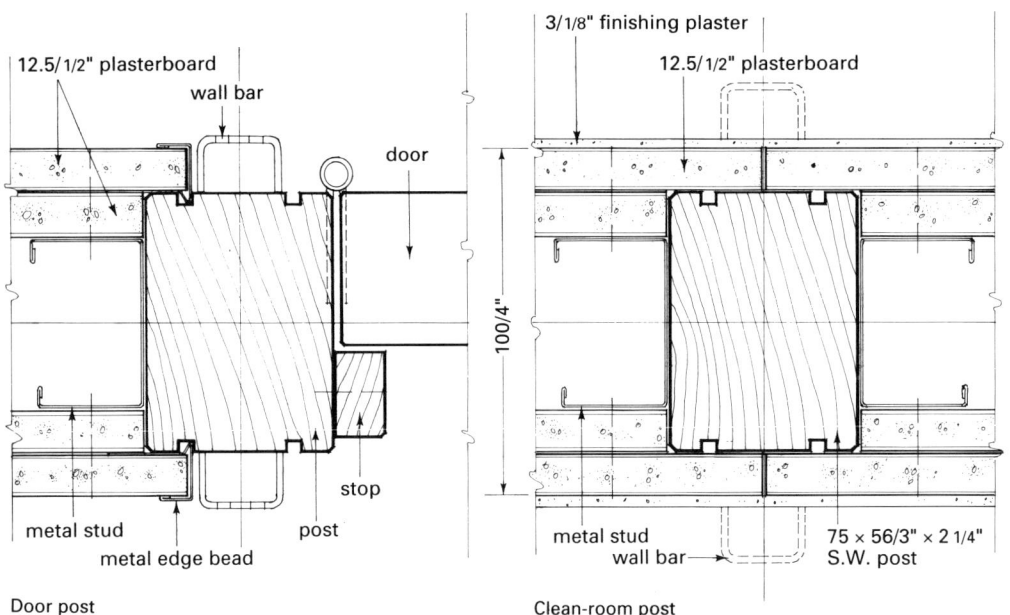

Door post

Clean-room post

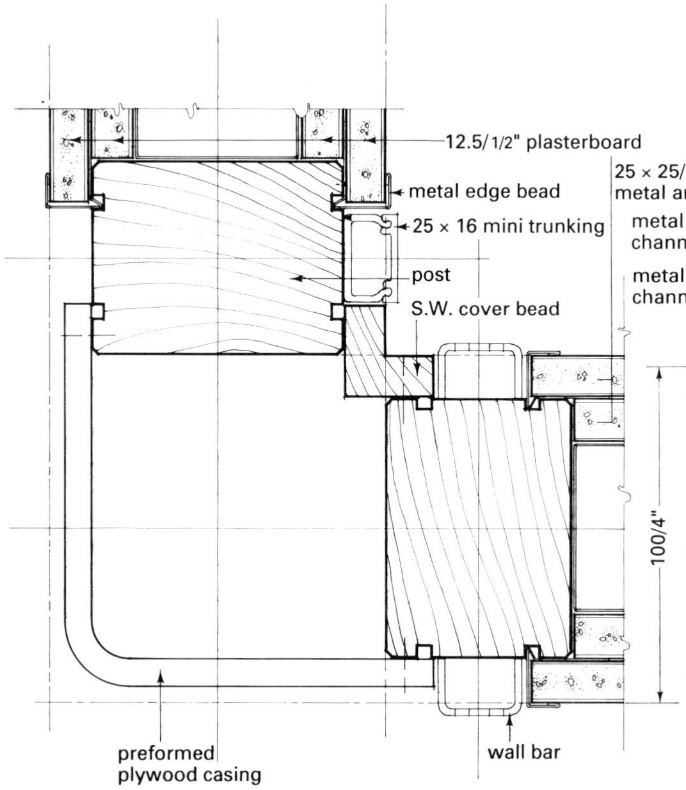

Junctions

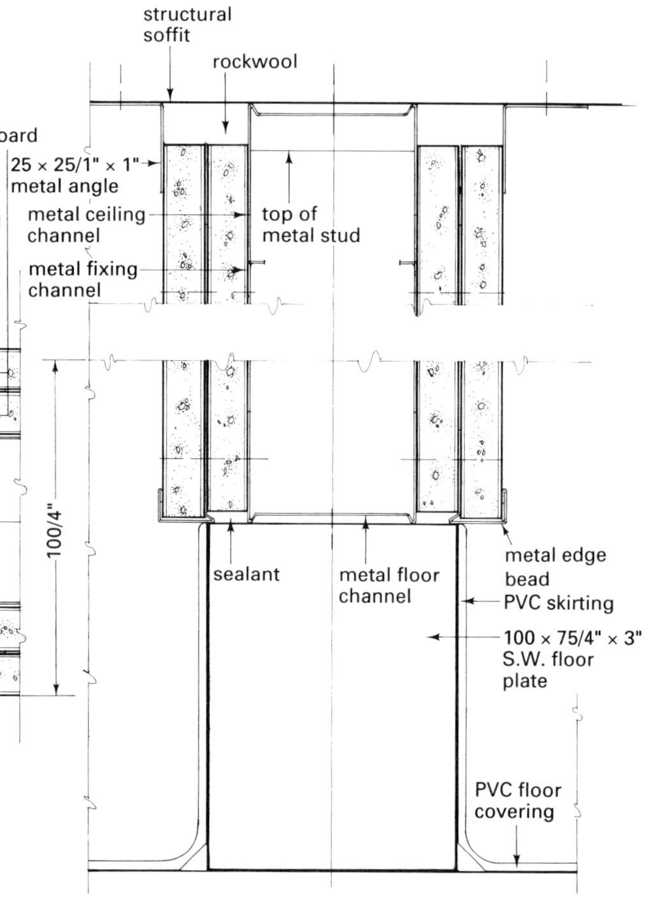

Sections

The building fabric

covered with fittings and equipment, which protect them from mechanical damage. Some non-standard labs and ancillary rooms requiring a high standard of cleanliness, or which will be washed down regularly, such as animal rooms, biohazardous labs and wash-ups, will require a tougher finish, e.g. gloss paint or a sprayed plastic. The latter is applied by specialists and the same problems as those mentioned for epoxy resin floors can arise.

6.3 Doors

A 900 mm (3 ft) wide leaf will allow access for most things that are taken into labs, but occasionally wider openings are needed for bulkier equipment. To meet this eventuality it is prudent to provide 1½ leaf doors, consisting of a leaf of about 900 mm (3 ft) wide for normal access together with a leaf between 300 mm (1 ft) and 500 mm (1 ft 8 in) wide, the latter normally kept shut and only opened when specifically required. The meeting styles should be rebated and the large leaf should have an observation panel. Lab door leaves should be solid and be fitted with mortice locks, lever handle furniture and kick plates, the half leaf with flush bolts. In the partition system described above, door leaves can be hung direct from the partition posts without additional door linings or frames. This will mean that leaf widths will derive from the space between posts and therefore from the module used. For example, a 1000 mm (3 ft 3 in) module will give a leaf width of 940 mm (3 ft 1 in) in a 1 module opening. 1½ leaf doors require a 1½ module opening, giving a 940 mm (3 ft 1 in) main leaf and a 500 mm (1 ft 8 in) secondary leaf.

Standard door widths cannot therefore be used, but standard heights and thicknesses can be. In practice, on all but the very smallest lab projects door leaves are not bought 'off the shelf' but are manufactured for the project, and in the UK the requirement for non-standard widths has not caused problems.

A painted finish is acceptable in most labs, but veneers and plastic laminates are also used, the latter being especially favoured for clean rooms.

Doors to animal rooms should have 1050 mm (3 ft 6 in) wide solid leaves (which may require an additional off-module support post), with leaves and frames clad in PVC for washing down. If fitted with observation panels these should be able to be obscured from both sides (a suitable type of blind in the UK is the Vistamatic shuttered vision panel by Steele's of Hainault, Essex).

6.4 Floor finishes

The standard floor covering for labs is sheet PVC with welded joints, welded to coved PVC skirtings to provide a flush, easily cleaned and waterproof floor. Any pipes or ducts that rise through the floor should be provided with square skirting boxes to take the coved skirtings, to maintain the waterproof capability. Animal rooms and other floors such as wash-ups, which are regularly washed down, are generally finished with jointless epoxy resin flooring together with a coved skirting of the same material. Many types of epoxy resin floors are now available; they are expensive and by no means trouble-free and it is important to specify a reputable brand. All are laid by specialists, who are frequently subcontractors to the suppliers, and it is not always possible to obtain a rapid return to site to rectify faults. These floors normally have floor drains and a type suitable for the floor covering should be specified (see Chapter 5, section 5.5).

The plant in plant rooms is a potential source of leaks, so the floor covering should be waterproof and taken up into coved skirtings. Asphalt is often used but epoxy resin is increasingly preferred. All plant is mounted on concrete plinths, which sit on top of the structural floor, and the skirtings should be taken up and over them to maintain the waterproof capability. All pipes, cables and ducts passing through the floor must be provided with concrete upstands to accept the skirtings.

6.5 Ceilings

Where ceilings are of plaster or concrete then the same considerations apply as for walls: vinyl emulsion to standard labs and gloss paint/sprayed plastic to animal rooms and other non-standard rooms.

In rooms in which there is a high concentration of fume cupboards relative to the floor area it will not be possible to supply make-up air through local diffusers without producing draughts and it will be necessary to supply the air from a plenum through perforated metal ceiling tiles. These tiles have an easily cleaned metal surface, and the non-perforated version is probably the most suitable type of ceiling for labs in which a suspended ceiling is necessary to screen services. The tiles are removable for access to the ceiling void.

In many ancillary rooms, except where moisture is present, mineral fibre ceiling tiles in an exposed lay-in grid provide an economical and acceptable ceiling.

6.6 External envelope

6.6.1 Window wall

There is often the need to take discharge ductwork of up to 250 mm (10 in) diameter direct through the window wall to outside, from safety cabinets, fume hoods, etc. The need may not be present in the initial brief, but can often arise during subsequent changes in use.

To cater for this need it is sensible to provide a fixed fanlight or panel above each main window, through which the ductwork can be taken without disturbing the latter. If there is a suspended ceiling then the panel can front the ceiling void.

6.6.2 Louvres

Because louvres form part of the external envelope they are best chosen and detailed by the architect in collaboration with the building services engineer, who will supply the final size of intake for each air-handling unit. These sizes will not be known at initial design stage, so it is prudent to allocate a fresh-air inlet zone in the plant room (see Chapter 2, section 2.2.17(c)) for a bank of intake louvres, to which the intake ductwork for each AHU can be connected. If, after final sizing of AHU intakes, there is unused space in this zone it can either be left louvred, as permanent ventilation to the plant room, or blanked off. Louvres should always be fitted with birdmesh.

6.7 Builder's work in connection (BWIC)

Engineering services items are usually shown on the building services engineer's drawings by symbol or diagrammatically, and it then becomes the architect's responsibility to illustrate how these items relate to the building fabric: for example, their precise positions; the sizes of holes required and how these are to be made good; how the item is fixed; and how the floor, wall or ceiling finishings are finished against it.

Because there are so many of these items in lab buildings there is a greater demand for this information than is usual; and some of it can only be prepared after the subcontractor responsible for supplying and installing the items has finalized his procurement programme and so knows the equipment that he or she will be using. The information need not always be provided in immaculately prepared drawings; the author's practice is to use A4 freehand drawings, which can be prepared rapidly, can be faxed to site and are perfectly adequate for the purpose.

If this information is not supplied to the builder then the architect will be in no position to demand the replacement of work that is technically acceptable, but visually unacceptable. Some typical examples where such details

Builder's work ducts

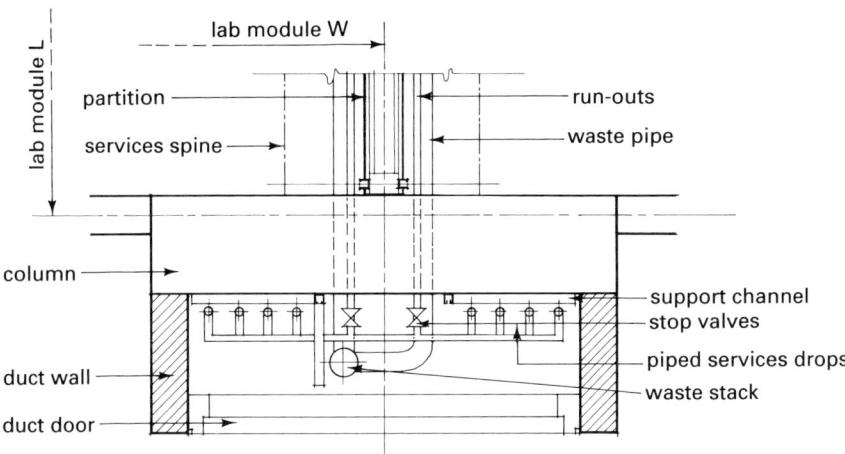

Option 1

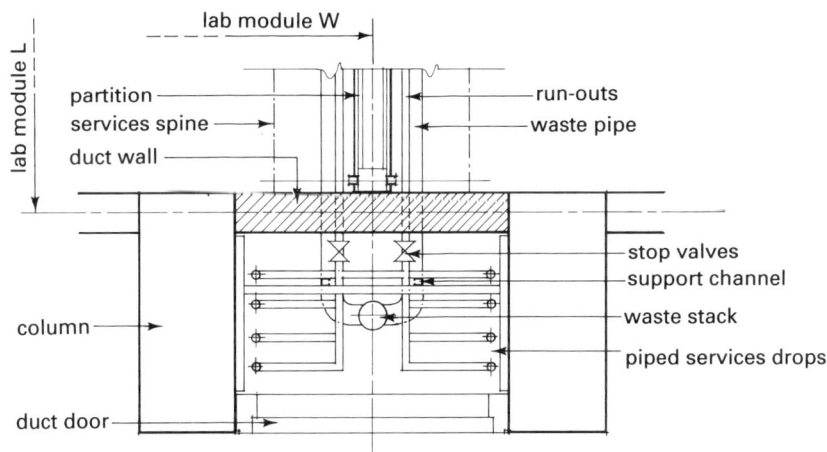

Option 2

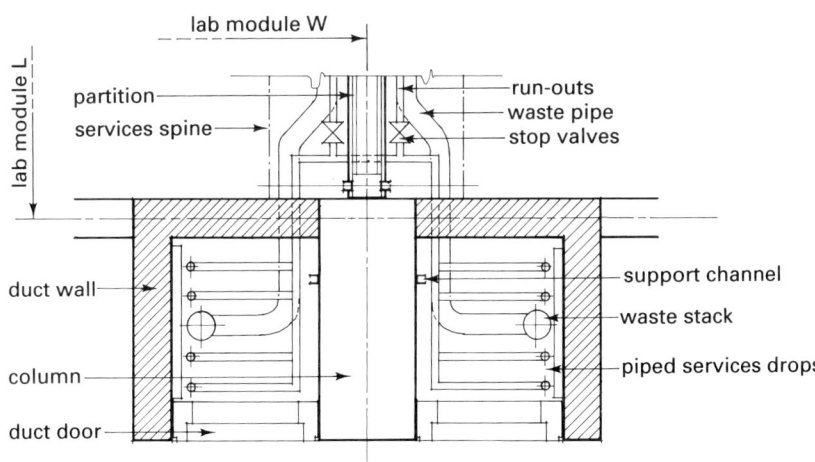

Option 3

may be required are:

- fire dampers to ductwork passing through floor slabs – those in the plant room showing the upstand required to maintain the waterproof characteristic of the floor (see section 6.4);
- pipes/cables passing through the plant room floor slab, showing upstands around them to maintain the waterproof characteristic;
- outlets and drains in floors, showing the height at which they are to be fixed, how their traps are accommodated in suspended floors, how floor coverings relate to them, etc.;
- fans in external walls (which in brick or masonry walls will require a timber lining);
- louvres (see section 6.6.2);
- diffusers and grilles in suspended ceilings;
- ductwork and piping exposed to view where it emerges from walls, partitions, floors or ceilings;
- pipe casings;
- Drenchshowers and eyewashes.

6.8 Builder's work ducts

The vertical ducts referred to in Chapter 1, section 1.5.3, are usually formed by the builder as part of the building fabric. Their positions will often coincide with those of the structural frame columns (see Chapter 2, section 2.1.5), and when this occurs difficulties can arise, because the column effectively divides the duct into two separate parts and complicates the routeing of the connections to the run-outs, as shown in the illustration.

One solution often adopted, which the author has used, is to replace the single column with a pair of widely spaced columns, which form the sides of the duct and free it from obstructions, simplifying the connections to run-outs, as shown in the illustration.

Another possibility is to turn the column through 90°, to form the back of the duct, although this decreases the useful depth of the duct (or increases its overall depth) and necessitates holes being cast in the column.

Ducts on the corridor side of the lab must be fitted with access doors of the same fire resistance as the duct.

A slot is normally formed through the structural floor slab to allow the sub-mains to penetrate; this must be filled in after the pipes are installed, usually with concrete, to maintain the fire resistance of the floor slabs.

6.9 Fire compartment penetrations

Where ventilation ductwork passes through a fire compartment boundary, whether a wall, partition or structural floor, it must be fitted with fire dampers to maintain the integrity of the compartment. The dampers have a mechanism for resetting after they have been actuated and access must be provided to allow this to be done.

Steel or cast-iron pipes passing through a fire compartment boundary are not normally considered to be a fire risk. In the UK, plastic pipes of over 50 mm (2 in) OD must be fitted with a collar consisting of a steel sleeve with an intumescent liner, which encloses the piping where it passes through the boundary. At a temperature of about 150 °C (300 °F) the intumescent liner expands inwards to close and seal the pipe. A suitable collar in the UK is the Firesleeve range produced by Dufaylite Developments Limited.

7 The design team

This chapter presents a brief introduction to the members of the design team, their responsibilities, what they do and how they are paid, and is primarily for those involved in appointing and/or briefing the team.

7.1 Members of the team

The design team responsible for the design of buildings normally consists of architect, structural engineer and building services engineer. In the UK and some Commonwealth countries a quantity surveyor is also a member of the team.

7.2 Responsibilities

7.2.1 The architect

Traditionally leads the design team and is the only member having an overview of the design of the whole project, in addition to being responsible for the design of the building itself and for coordinating the contributions of the structural engineer and the building services engineer. In the UK the architect administers the building contract and issues the monthly certificates for payment. The architect's fees are usually a percentage of the total cost of the building contract.

7.2.2 The structural engineer

Is responsible for the structural frame of the building: that is, for its supporting elements, such as the foundations, columns and floor slabs. The structural engineer's fees will generally be a percentage of the cost of the structure.

7.2.3 The building services engineer

Is responsible for the design of the engineering services, such as power, lighting, water supplies, drainage, special gases, mechanical ventilation and air conditioning, and also for advising on the costs of these services. The building services engineer's fees will generally be a percentage of the cost of the engineering services installations.

7.2.4 The quantity surveyor

Where the quantity surveyor is a member of the design team he or she will advise on the cost of design solutions proposed for the building and its structural frame by architect and structural engineer during the design stages, will receive the building services engineer's estimated cost of the services, and will prepare an estimate of cost (for the whole project) of the design team's solution. This will provide the basis on which the client decides whether to proceed with the scheme, or whether economies must be made. The quantity surveyor prepares the bills of quantities (which are the main tender document where a quantity surveyor is employed) and, during the construction period, prepares a monthly valuation of the value of the completed work on which the architect issues a certificate for payment. The quantity surveyor's fees are a percentage of the total cost of the building contract.

7.3 How the team functions

The design of a laboratory building is very much a team effort, so it is important for each member of the design team to be able to make contributions as early as possible; for this reason the client should be encouraged to appoint the whole of the design team at the inception of the project.

Members' contributions should not be confined to their particular discipline. For example, in formulating initial proposals for the design concept the architect will need to have his or her own 'in principle' proposals for the structural system, the method of supplying the bench outlets, the policy for removing fume cupboard fumes, on supplying make-up air and on the general services strategy, in order to provide a basis for discussion with the other team members.

These 'in principle' proposals will show the other members of the team the architect's overall concept and will enable them to approve the proposals or to make positive suggestions either for modifying them or for replacing them with more effective means of attaining the required result.

The contribution of members to disciplines other than their own can often act as a catalyst in producing fresh ideas in the member responsible for that discipline.

The architect normally leads the design team because he or she is responsible for the conceptual design of the project as a whole, and not solely for the building fabric.

The structural frame solution that the architect agrees with the structural engineer will be a solution that in the architect's judgement is best suited to the requirements of that particular building and which can be provided within the budget constraints of the project.

The engineering services solutions that the architect agrees with the building services engineer will be solutions that in the architect's judgement are best suited to the project and which can be provided within the budget constraints of the project.

The architect's opinion will therefore act as the casting vote in the type of structural frame to be used and in the 'in principle' servicing solutions. But of course no structural engineer or building services engineer worth their salt will agree to solutions that, as the designers ultimately responsible for their specialities, they cannot support.

Where the quantity surveyor is a member of the team, the architect will be consulting the quantity surveyor on the cost effect of the various proposed designs, for the building fabric as well as for the structure and services.

7.4 Programmes

An outline programme will normally form part of the design team's proposals to the client. It is usual for this to be prepared by the architect, with input from the structural engineer and the building services engineer. Construction times have a direct bearing on costs, so the quantity surveyor will also have an input in preparing the programme.

7.5 Design team meetings

Discussions and consultations will take place at design team meetings, which the architect normally arranges, chairs and minutes. The frequency of these will depend upon the nature of the project, but monthly meetings are usual. In between the formal meetings there will of course be informal meetings between team members, at which design proposals and options will be canvassed, discussed and agreed.

Technical supplements

88	TS1	Radioactive laboratory classification
89	TS2	Containment of dangerous pathogens
90	TS3	Laboratory containment facilities for genetic manipulation
91	TS4	Microbiological safety cabinets
92	TS5	Clean rooms
94	TS6	Initial questionnaire for laboratory projects
95	TS7	Room data sheet
96	TS8	Laboratory furniture performance specification
100	TS9	Laboratory furniture menu
107	TS10	Laboratory furniture suppliers
108	TS11	Fume cupboard criteria
109	TS12	Fume cupboard schedule
110	TS13	Fume cupboard performance specification
113	TS14	Schedule of taps and valves
114	TS15	Bench outlets: colour code chart
115	TS16	Services into each laboratory module
116	TS17	A servicing concept for research laboratories
120	TS18	Chemical resistance chart
123	TS19	Radioactive shielding
124	TS20	Fire extinguishers
125	TS21	Scientific disciplines
127	TS22	External design temperatures
130	TS23	Conversion data

TS1 Radioactive laboratory classification

(From *Symposium on Design of Laboratories for Work with Radioactive Materials*; in ascending order of sophistication for standards of finish, fittings and facilities)

Grade C laboratory

As a rule most work will be with weak active solutions in glass vessels, placed in trays to provide double containment. Most new laboratories will be suitable for this classification: for example, welded sheet PVC flooring and laminate-covered worktops, without non-standard features being required. For the simplest work no fume cupboard will be required but if there is a possibility of radioactive gases, vapours or particulates being released then an efficient fume cupboard will be essential. Air extracted from the fume cupboard need not be filtered.

Grade B laboratory

This is the usual classification envisaged when labs are described as 'radioactive'.

This grade of lab is for work with moderate amounts of activity including many procedures for preparing labelled compounds. It is best provided as a completely new building, though a conversion may be possible. The design must take account of the larger amount of activity so as to keep contamination levels under control, and to facilitate decontamination should a major spill occur (i.e. one where a substantial part of the total activity in the laboratory is dispersed). There will be routine use of efficient fume cupboards. At the upper end of the scale of usage overshoes and a barrier at the entrance will usually be provided, reached via a properly equipped changing room. At the other end of the scale the barrier will be more by way of example.

All surfaces – walls, ceiling, floor, laboratory furniture and fittings – must be smooth and impervious, and all gaps sealed. Adequate cupboard space to keep reagents and apparatus not actually in use is essential. Elbow-operated taps are to be preferred, both in the laboratory and the changing room. Filtration of air from the exhaust ventilation system will not usually be necessary but should be reviewed at the design stage.

Grade A laboratory

These must be carefully planned as such prior to building. They are for work with the more toxic radioisotopes (mainly Class 1) at other than trace levels. Mostly the work has to be done under totally enclosed conditions, i.e. in glove boxes. The highest standard of finish is necessary, much use being made of stainless steel for working surfaces, all joints carefully sealed and meeting surfaces coved. A plenum system of ventilation would be essential. All air from exhaust ventilation systems must be well filtered: 'absolute' filters are commonly used. Access would be by a fully equipped change room and barrier, and would be rigidly controlled.

TS2 Containment of dangerous pathogens

Hazard groups

Group 1 An organism that is most unlikely to cause human disease.

Group 2 An organism that may cause human disease and which might be a hazard to laboratory workers but is unlikely to spread in the community. Laboratory exposure rarely produces infection, and effective prophylaxis (preventive treatment) or effective treatment are usually available.

Group 3 An organism that may cause severe human disease and present a serious hazard to laboratory workers. It may present a risk of spread in the community but there is usually effective prophylaxis or treatment available.

Group 4 An organism that causes severe human disease and is a serious threat to laboratory workers. It may present a high risk of spread in the community and there is usually no effective prophylaxis or treatment.

Categories of containment: laboratory containment levels

Tabular summary of laboratory containment requirements

Containment requirements	Containment levels			
	1	2	3	4
Laboratory site: isolation	No	No	Partial	Yes
Laboratory: sealable for fumigation	No	No	Yes	Yes
Ventilation:				
inward airflow/negative pressure	Optional	Optional	Yes	Yes
through safety cabinet	No	Optional	Optional	No
mechanical: direct	No	No	Optional	No
mechanical: independent ducting	No	No	Optional	Yes
Airlock:	No	No	Optional	Yes
with shower	No	No	No	Yes
Wash handbasin	Yes	Yes	Yes	Yes
Effluent treatment	No	No	No	Yes
Autoclave site:				
on site	No	No	No	No
in suite	No	Yes	Yes	No
in lab: free-standing	No	No	Optional	No
in lab: double-ended	No	No	No	Yes
Microbiological safety cabinet/enclosure	No	Optional	Yes	Yes
Class of cabinet/enclosure	–	1	1/3	3

Radioactive laboratory classifications

TS3 Laboratory containment facilities for genetic manipulation
(from Advisory Committee on Genetic Manipulation Note 8, 1988)

Tabular summary of laboratory containment requirements

Containment requirements	Containment levels			
	1	2	3	4
Laboratory site: isolation	No	No	Partial	Yes
Laboratory: sealable for fumigation	No	No	Yes	Yes
Ventilation:				
inward airflow/negative pressure	Optional	Optional	Yes	Yes
through safety cabinet	No	Optional	Optional	No
mechanical: direct	No	No	Optional	No
mechanical: indepedent ducting	No	No	Optional	Yes
Airlock:	No	No	Optional	Yes
with shower	No	No	No	Yes
Wash handbasin	Yes	Yes	Yes	Yes
Effluent treatment	No	No	No	Yes
Autoclave site:				
on site	No	No	No	No
in suite	No	Yes	Yes	No
in lab: free-standing	No	No	Optional	No
in lab: double-ended	No	No	No	Yes
Microbiological safety cabinet/enclosure	No	Optional	Yes	Yes
Class of cabinet/enclosure	–	Class I	Class I/III	Class III

TS4 Microbiological safety cabinets (from BS 5726 : 1992)

These cabinets are principally designed to provide some protection for the users and the environment from the hazards associated with handling dangerous biological material. The techniques employed in the manipulation of microbiological organisms can produce aerosols, which can be readily inhaled by laboratory scientists and inadvertently distributed to the surrounding environment. Dangerous pathogens are grouped into four categories of hazard (Technical supplement 2) and there are three classes of biological safety cabinet, each designed to cater for a degree of hazard. The choice of cabinet appropriate to the degree of hazard will be the responsibility of the client, and the recommendations below are given merely as a guide.

Class 1
An open-fronted, exhaust-protected cabinet, through which the operator can carry out manipulations inside the work area. Room air flowing into the cabinet minimizes the escape of aerosols from the work area and is exhausted to the outside through a high-efficiency (HEPA) filter system. Generally recommended for hazard groups 1–3.

Class 2
An open-fronted cabinet in which the work area is flushed with a unidirectional downward airflow. An inward flow of air from the lab forms an air curtain, which minimizes the escape of aerosols from the work area, while the filtered air within the cabinet protects the work from airborne contamination. All the contained air may be recirculated, or a proportion may be recirculated with the remainder exhausted to the outside through a high-efficiency (HEPA) filter. These are often referred to as laminar-flow biological safety cabinets. Generally recommended for hazard groups 1 and 2.

Class 3
A totally enclosed, gas-tight, ventilated cabinet where the operator is physically separated from the work area. The work is conducted through rubber gloves attached to the cabinet. Room air is drawn into the cabinet through a high-efficiency (HEPA) filter, which prevents the escape of aerosols from the cabinet as well as supplying the work area with sterile air. The exhaust air to the outside is also through a high-efficiency (HEPA) filter system and when the cabinet is in use it is maintained at a negative pressure. Generally recommended for hazard groups 1–4.

TS5 Clean rooms

1 Standards UK – BS 5295 : Part 1, 2 & 3: 1989 Clean Rooms, Work Stations and Clean Air Devices

USA – US Federal Standard 209

2 BS 5295 data

2.1 Guidance on selection of classes of clean rooms

- *Class 1* – used when an ultra-clean bacteria-free and particulate-free atmosphere is required; for example, in certain stages in the manufacture of some injectable medicinal products, the assembly and test of microelectronic units and exceptionally sensitive mechanisms, and for surgical operations.
- *Class 2* – used for the assembly and test of precise gyroscopes, precision bearings, solid-state electronic devices, certain high-quality optical equipment, very close-tolerance hydraulic and pneumatic devices, and for the manufacture of some medicinal products. Where the amount of work requiring this level of cleanliness is small, the use of contained workstations inside an area of lesser cleanliness should be considered.
- *Class 3* – used for the assembly of precision hydraulic and pneumatic system subassemblies, servo-control valves, electromechanical and precision timing devices, certain optical instruments, small roller and ball bearings, and high-grade gearing manufacture.
- *Class 4* – used for general optical work, the assembly and test of electronic assemblies and measuring equipment, and for the assembly of those hydraulic, pneumatic and lubrication components, including pumps and motors, where the use of a controlled area does not provide a sufficiently high order of cleanliness.

2.2 Summary of requirements

Controlled environment (clean room, workstation or device)	Recommended airflow configurations	Recommended periodicity for air sampling and particle counting	Max. permitted number of particles per m^3 (equal to, or greater than, stated size)					Final filter efficiency %
			$0.5\mu m$	$1\mu m$	$5\mu m$	$10\mu m$	$25\mu m$	
Class 1	Unidirectional	Daily or continuous by automatic equipment	3 000*	Not applicable	Nil	Nil	Nil	99.995
Class 2	Unidirectional	Weekly	300 000*	Not applicable	2 000	300	Nil	99.95
Class 3	Unidirectional or conventional	Monthly		1 000 000	20 000	4 000	300	95.00
Class 4	Conventional	3-monthly			200 000	40 000	4 000	70.00
Controlled area	Normal ventilation	–	–	–	–	–	–	–
Contained workstation	Unidirectional	To suit required class application						99.997
Portable tent	As selected	To suit required class and application			To suit required class		To suit required class	To suit required class

(*Subject to maximum particle size of $5\mu m$.)

2.3 Design and construction

1. The size of the room should be kept to the minimum practicable.
2. Normal (non-emergency) access to and egress from the clean room should be through airlocks for both material and personnel. Airlocks may be used as anterooms.
3. Air-showers, step-over benches and similar decontamination devices and procedures within the airlock system should be employed as appropriate.
4. Anterooms should be provided with washing and toilet facilities, stowage facilities for clothes and changing room space and should be designed in three distinct sections: locker, partially contaminated, uncontaminated.
5. The use of small pass-through devices should be considered where small items are involved, from the airlock either to the clean room or to a lobby.
6. Pairs of airlock doors should be interlocked to prevent both being opened simultaneously.
7. Windows on outside walls should be avoided wherever possible in order to reduce heat loss, condensation and noise problems.
8. Windows should be non-opening, flush-fitting and sealed to prevent ingress of contamination.
9. The design should exclude all unnecessary ledges and similar surfaces where contamination can accumulate.
10. All fittings should be flush and easily cleanable; only those that are essential to the operations being carried out are to be fitted within the room. All other fittings, such as fuse boxes, switch panels, isolators and valves, should be located outside.
11. All internal surfaces should be smooth and free from cracks, ledges and cavities. Corners should be reduced and all permanent pipes and cables suitably boxed in with low particle-shedding materials.
12. Floor coverings should be continuous; where sheet materials are used the joints must be welded flush.
13. Lamp fittings should be flush with the ceiling or wall, totally enclosed and wherever possible provided with external access for maintenance.
14. The use of overhead electrical sockets should be considered.
15. Piped liquids and gases should be filtered before entering the clean room to ensure that the liquid or gas at the work position will be as clean as, or cleaner than, the air circulating at that point. Non-ferrous or stainless steel piping and fittings should be used.
16. Personnel working in clean rooms should wear protective garments, designed to prevent contamination generated by the body and retained on everyday clothing from being transmitted into the room. These should comprise headcover, smock and trousers (or one-piece overall suit), shoes or overshoes.

TS6 Initial questionnaire for laboratory projects

1. Client _____

2. Name of project _____

3. Description _____

4. Site _____

5. Type: Teaching ☐   Routine ☐   Research ☐

6. Discipline: Chemistry ☐   Physics ☐   Biology ☐
   Other _____

7. Radioactive ☐
   Grade: C ☐   B ☐   A ☐

8. Biohazardous ☐
   Containment level: 1 ☐   2 ☐   3 ☐   4 ☐

9. Bench services position: On bench ☐   Services spine ☐   Services bridge ☐

10. Lab furniture
    Bench support system: Pedestal units ☐   Table frame ☐   Cantilever ☐
    Benchtops: Timber ☐   Laminate on core ☐   Solid laminate ☐   Epoxy resin ☐
    Other _____
    Underbench units: Pedestal ☐   Suspended ☐

11. Ventilation: Natural ☐   Mechanical ☐   Cooling ☐   A/C ☐

12. Refurbishment projects
    Complete strip-out within site boundary ☐
    New room layout(s) ☐
    New services spines/run-outs/outlets ☐
    New lab furniture ☐
    Existing lab furniture reused ☐
    Existing room layout(s) to be retained ☐
    Existing services spines to be retained ☐
    Existing run-outs/outlets to be upgraded ☐
    Existing mechanical ventilation to be upgraded ☐
    New mechanical ventilation installation ☐
    New fume cupboard(s) ☐
    Existing fume cupboard(s) reused ☐

TS7 Room data sheet

01 Function
Room number: _____ Occupancy: _____
Room description: _____
Time of use: regular _____ to _____ occasional _____ to _____

02 Physical criteria Area: _____
Ceiling height: ☐ normal ☐ special _____ minimum
Loading: ☐ normal ☐ heavy _____

03 Environmental criteria
Temperature: ☐ normal ☐ special _____ to _____ °C
Humidity: ☐ normal ☐ special _____ to _____ %RH
Ventilation: Natural: ☐
    Mechanical: ☐ normal ☐ special _____
    Positive pressure: ☐ Negative pressure: ☐
Smoke, fumes, odours: ☐ _____
Daylight: ☐ undesirable ☐ desirable ☐ essential
Artificial lighting: Level: ☐ normal ☐ high _____
    Variable: ☐ Timed: ☐
    Safelights: ☐ Ultraviolet: ☐
Noise: Containment: ☐ no problem ☐ problem _____
    Exclusion: ☐ normal ☐ critical _____
    Vibration isolation: ☐ normal ☐ critical _____

04 Safety/security
Fume cupboard(s): ☐
Microbiological safety cabinet(s): ☐ _____
Toxic material: ☐ _____
Radiation: ☐ _____
Fire/explosion risk: ☐ _____
Electrical screening: ☐ _____
Drenchshower: ☐ Eyewash: ☐ _____

05 Building fabric
Floor finishes: ☐ vinyl ☐ carpet ☐ other _____
Wall finishes: ☐ normal ☐ washable ☐ other _____
Ceiling finishes: ☐ normal ☐ washable ☐ other _____
Windows: ☐ curtains ☐ blind ☐ blackout blind
Doors: ☐ single ☐ 1½ ☐ double
    ☐ glazed ☐ ½ glazed ☐ observation panel
    ☐ latch ☐ lock ☐ keycard access

06 Piped services
Cold water: ☐ pressure _____
Drinking water: ☐ Hot water: ☐
Purified water: ☐ RO ☐ deionized ☐ distilled
Fuel gas: ☐ Vacuum: ☐
Compressed air: ☐ pressure _____
Acetylene: ☐ Steam: ☐
Nitrogen: ☐ Oxygen: ☐
Carbon dioxide: ☐ Carbogen: ☐
Nitrous oxide: ☐ Helium: ☐
Others: _____

07 Drainage
Sink(s): ☐ Dripcups: ☐ Floor drain: ☐
Abnormal fluids: ☐ solvents ☐ acids ☐ alkali ☐ radioactive
Other: _____

08 Electrical
Service: ☐ normal ☐ three-phase ☐ dedicated
Power: ☐ 13 A sockets ☐ 20 A/30 A ☐ low-voltage ☐ DC ☐ waterproof
    ☐ sparkless
Dedicated earth: ☐
Other: _____

09 Communications: Telephone: ☐ Data: ☐ Clock: ☐

10 Fittings/fixtures
Benching: ☐ Underbench units: ☐ Wall cupboards: ☐ Shelves: ☐
Writing board: ☐ Pinboard: ☐ Coathooks: ☐ Mirror: ☐
Other: _____

11 Other requirements

TS8 Laboratory furniture performance specification

Note: This specification assumes a 1000 mm (3 ft 3 in) furniture module. When other modules are used the dimensions should be altered to suit.

1 Generally

1.1 Scope of contract

The contract includes the whole of the work shown on the drawings and described in this specification.

The successful tenderer will be required to become a nominated subcontractor to the main contractor under the form of tender named in the invitation to tender letter.

Tenders shall include for arranging for deliveries to site to suit the main contractor.

1.2 Included in the work

Supply, placing and bolting in position of all benches, including those with inset sink bowls ('lab sinks').

Supply and fixing of sink units.

Supply and placing in position of underbench cupboard/drawer units, suspended from bench frames.

Supply and fixing to walls and posts of wall bars complete with brackets to support wall cupboards and shelves.

Supply and fixing on brackets of wall cupboards and shelves.

1.3 Excluded from the work

All services spines, service outlets and dripcups.

All piped services, traps and wastes, including final connections to waste tails of sink bowls.

All electrical trunking, conduit and wiring.

1.4 Specification of components

The components are to a 1000 mm module, i.e. wall bars are at 1000 mm centres, wall cupboards and shelves are 1000 mm long, benches are 1000 mm, 1500 mm and 2000 mm long, and underbench units are ½ module and 1 module long.

The descriptions of the components that follow are in outline only and tenderers may offer their standard materials, construction and finishes where these comply with the size, performance requirement and finish called for.

Where finishes are specifically stated these indicate a performance requirement and shall not be varied.

Tenders must include details of the appearance, construction, finishes, hinges and handles of the components offered.

The successful tenderer may be required to provide a sample of each component for approval before commencing production.

The furniture will be installed in a centrally heated building.

1.5 Drawings

The tender drawings will be listed in the invitation to tender letter and will include general arrangement plans showing the location of the components together with details of each component.

2 Components

2.1 Benches

Shall consist of table frames supporting tops, and will generally be 600 mm deep and 1000 mm, 1500 mm or 2000 mm long, and will generally be available in two heights, 900 mm (H) or 720 mm (L).

Bench frames shall have square, minimum 30 mm, tubular steel legs fitted with approved adjustable feet with screw height-adjustment, connected at top with rails minimum 40 mm high, suitable for supporting the underbench units, and connected approx. 150–200 mm above floor along one side and at both ends with similar rails. The legs and rails shall be rigidly connected either by welding or other approved means to provide a rigid, stable and accurate assembly. Benches 2000 mm long shall be provided with an intermediate pair of legs. All welds shall be ground flush and the frames coated in dark grey epoxy powder.

Provision shall be made in the back and side top rails for bolting benches together and to service ledges and spines.

Tops will generally be available in either standard or radioactive form and shall be securely fixed down to the steel frames with self-tapping screws from below.

Standard tops will be in 20 mm thick solid laminate, acid-resistant laboratory grade, with brown core and top layer decorative laminate of approved colour, as Trespa Valkern grade L or equal approved, matt finish. On bench codes suffix A, the tops shall project beyond the bench frames on one side to fill the spaces shown and shall be cut on site to fit.

Radioactive tops will be in cast epoxy resin of approved colour, with integral upstands on all four edges to form a tray, as produced (in the UK) by Messrs Simmons (Mouldings) Ltd, Parkside, Coventry or equal approved.

2.2 Lab sinks

Shall consist of sink bowls of the size and material shown on the drawings, in cut-outs in benchtops, with joint between benchtop and sink flange sealed with silicone adhesive.

Bowls shall be complete with 38 mm Vulcathene 504 waste with tail for connection to trap by others.

Stainless steel bowls shall be in 18/8 316 grade acid-resisting stainless steel.

Epoxy resin bowls shall be as supplied by the radioactive top supplier.

Polypropylene bowls shall be Vulcathene.

2.3 Underbench units

Underbench units shall be without plinths and shall be suspended from the front and back top rails of benches in a manner that will allow the unit to be moved along the length of the bench or removed completely.

They shall be nominally 470 mm or 940 mm long (so that two 470 mm units or one 940 mm unit will fit neatly between the legs of a 1000 mm bench). They shall be approximately 500 mm deep, 630–700 mm high for 900 mm high benches and 450–520 mm for 720 mm high benches.

They shall consist of cupboards or drawers, or both, as on the drawings.

Carcasses shall have sides, top and bottom minimum 18 mm thick, chipboard or medium-density fibreboard, with 6–7 mm plywood/MDF back, all rigidly jointed to prevent distortion.

Backs shall be finished in pigmented acid-catalysed lacquer. Sides, top and bottom shall be finished in veneer, decorative laminate or pigmented acid-catalysed lacquer, as specified. Where a veneer is specified this shall include solid edging to front edges in the same timber, and finishing all timber surfaces in 2 coats acid-catalysed lacquer. Where decorative laminate is specified this shall include edging the front edges in the same material.

Cupboards shall be fitted with one adjustable shelf of chipboard not less than 18 mm thick, covered with white laminate on all surfaces.

Drawer boxes shall be of solid timber or plywood construction properly jointed (thermoplastic drawer systems will not be accepted), sliding on strong runners. Their interiors shall be finished with one coat of sealing lacquer.

Doors shall be chipboard/MDF minimum 16 mm thick, finished in veneer or decorative laminate as specified. A veneer shall include solid edging in the same timber, and finishing all timber surfaces in 2 coats acid-catalysed lacquer. Decorative laminate shall be 0.8 mm thick of approved colour, edged in hardwood or matching laminate as specified. Doors shall be face-fixed 'lay on' type, on approved self-closing cabinet hinges and fitted with approved handles. Where specified they shall be fitted with approved cupboard locks.

Drawer fronts shall be of the same construction and finish as doors and fitted with the same handles.

Waste bin units shall have a bucket shelf and be complete with removable plastic bucket/bin.

Solvent storage cupboards shall be steel-lined and completed with retaining bund, single shelf and door locks and shall comply with the requirements of the Highly Flammable Liquids and Liquid Petroleum Gas Regulations 1972.

2.4 Wall bars

Shall be Spur single-sided uprights, epoxy powder coated to approved colour, of the lengths specified. They shall be screw-fixed into grooves in timber posts (posts by others) or plugged and screwed to walls, at 1000 mm centres. Alternative proprietary wall bars will not be accepted.

2.5 Wall cupboards

Shall be of a nominal length of 1000 mm to suit the centres of wall bars, and generally 300–350 mm deep and 700 mm high.

Carcasses shall be as underbench units but fitted with a stiffening rail along the front of the bottom to prevent sagging.

Cupboards shall generally be fitted with one adjustable shelf, as underbench units.

Glass sliding doors shall be 6 mm clear float glass, GG quality, with edges rounded, ground and polished, running in black PVC channel tracks and with approved finger pulls.

Solid doors shall be as underbench units.

Each wall cupboard shall be supported on and include 4 Spur Cabinet Brackets 9076, fixed to back of cupboard and slotted into wall bars.

2.6 Shelves

Shall be a nominal length of 1000 mm to suit centres of wall bars, supported on and including 2 Spur Shelf Ends 9073, colour to match wall bars; screwed to shelf and slotted into wall bars.

Reagent shelves shall be of 20 mm thick solid laminate as standard benchtops, of approved colour, 250 mm deep. Standard laminate covered shelves shall be of 25 mm thick five-ply blockboard with core strips parallel with long sides and covered on both faces with 0.8 mm white laminate securely bonded to core, edged in hardwood.

2.7 Sink units	Shall comprise sink tops of bowl(s) and drainer(s) mounted on standard leg frames as for benches, 900 mm high.
	Shall be of 18/8 316 acid-resisting grade stainless steel, minimum 1.2 mm (18 gauge) thick, bowls and drainers pressed in one piece, drainers fluted to fall unless otherwise stated, without tiling upstand at back unless stated and complete with tapholes, overflow and wastehole. Sinks shall be provided with sound-deadening material on the underside. Sink bowls shall be fitted with 38 mm Vulcathene 504 waste, 508 plug and chain assemblies and 509 overflow assembly, for connection to trap by others. Traps and taps will be supplied and fitted by others.
2.8 Skirtings	Shall be 18 mm thick × 50 mm high Trespa Matt surface, of approved colours, fixed with plastic-capped screws.
2.9 Sealing	Seal joints between benchtops, benchtops and skirtings, and benchtops and service spines (spines by others) with white silicone sealant, Dow Corning or equal.

TS9 Laboratory furniture menu

Lab furniture: elevation of components

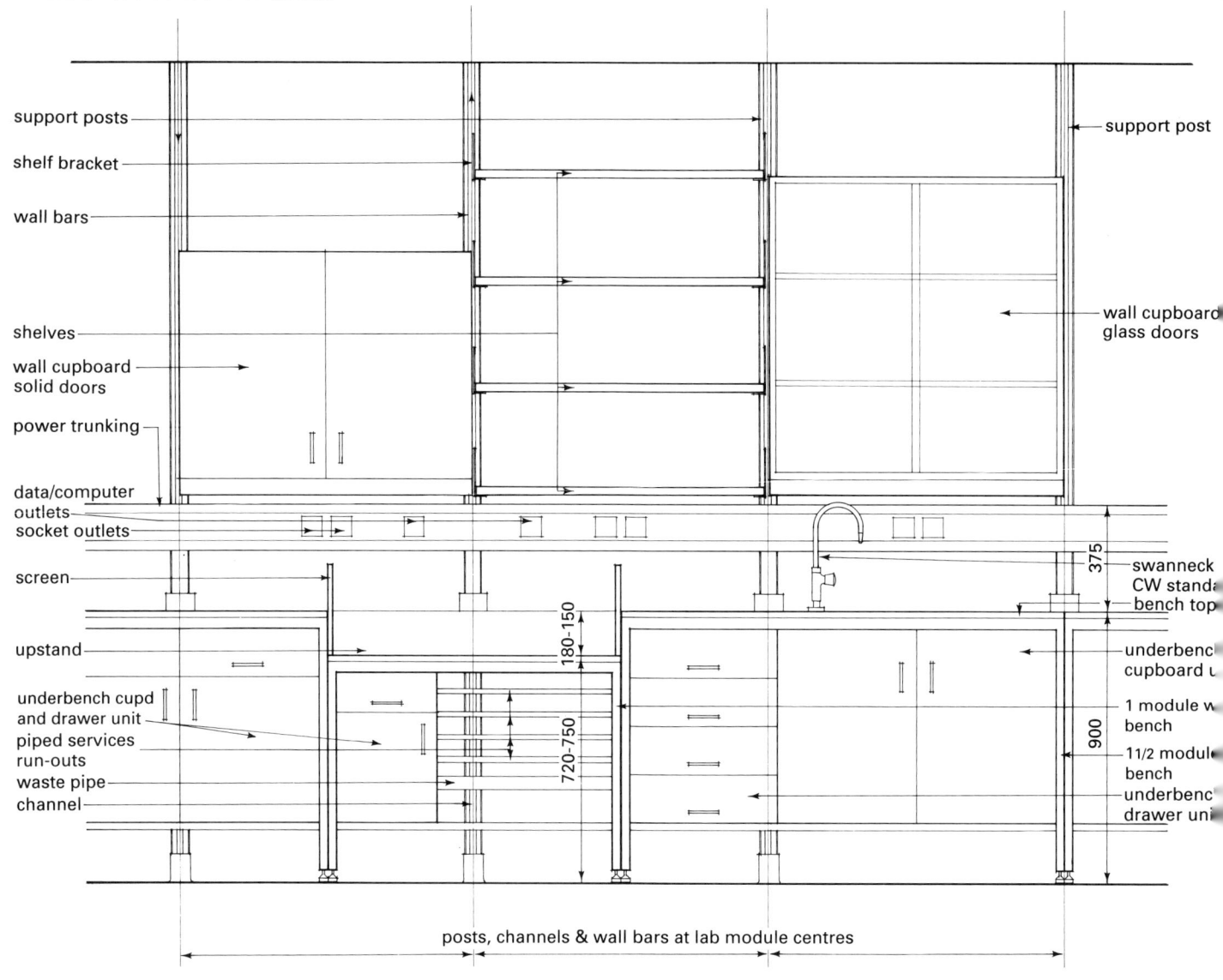

Lab furniture: sections

Peninsular/island

Wall

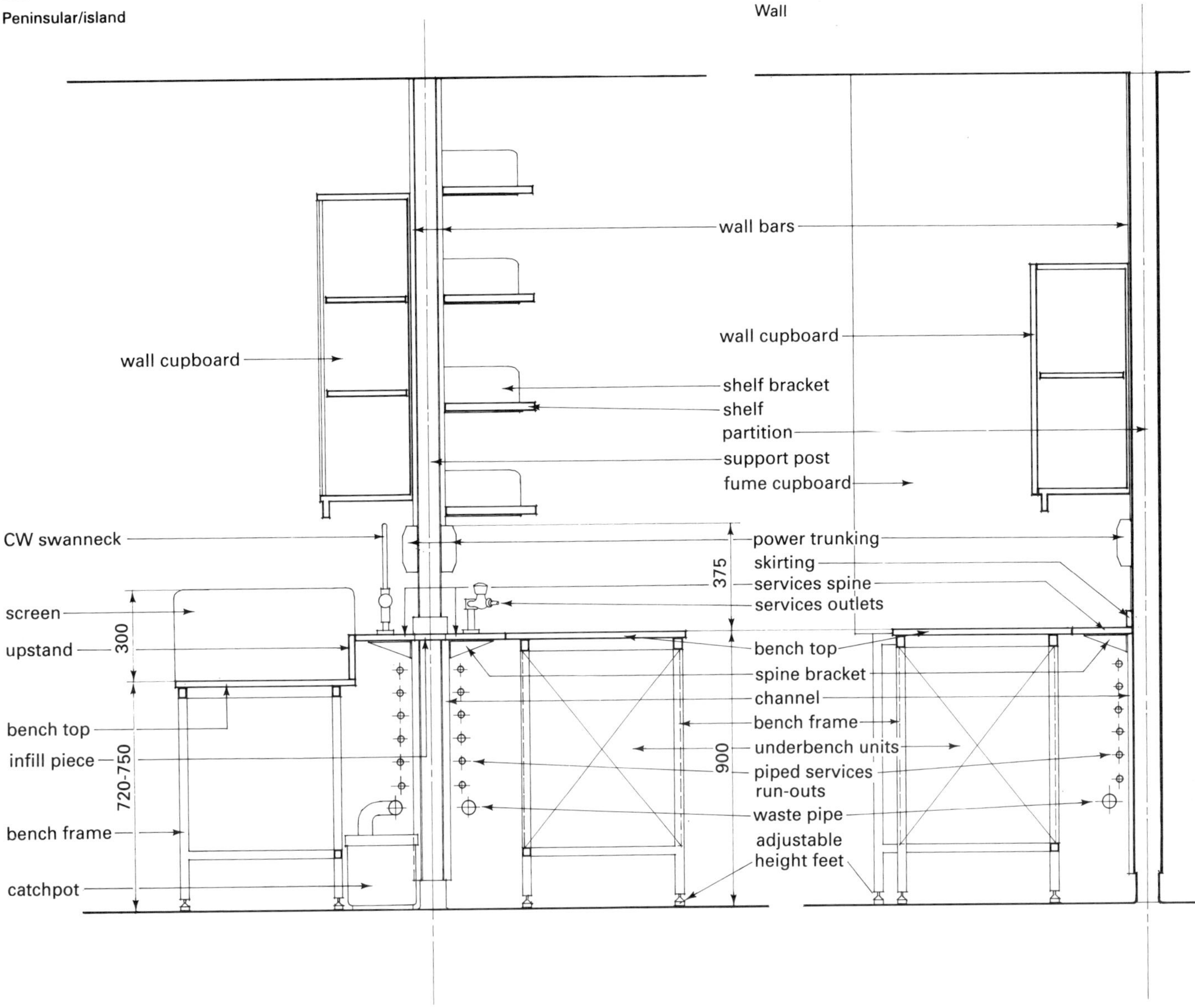

- wall bars
- wall cupboard
- wall cupboard
- shelf bracket
- shelf
- partition
- support post
- fume cupboard
- CW swanneck
- power trunking
- skirting
- services spine
- services outlets
- screen
- upstand
- 300
- 375
- bench top
- spine bracket
- channel
- bench top
- infill piece
- bench frame
- 720-750
- underbench units
- 900
- piped services run-outs
- waste pipe
- bench frame
- adjustable height feet
- catchpot

Laboratory furniture menu

101

Lab furniture: standard benches range

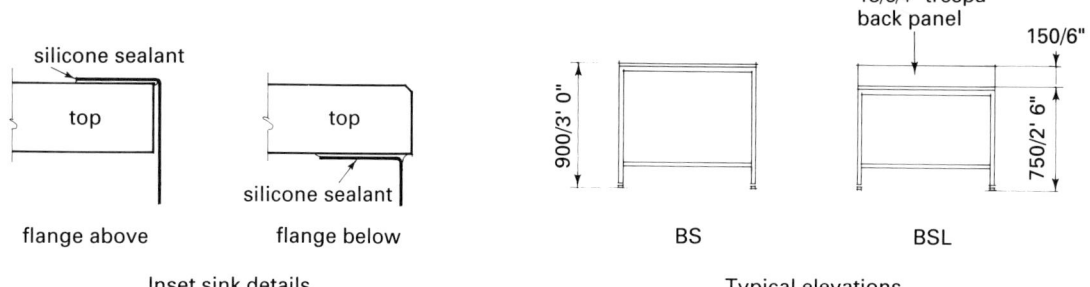

Inset sink details   Typical elevations

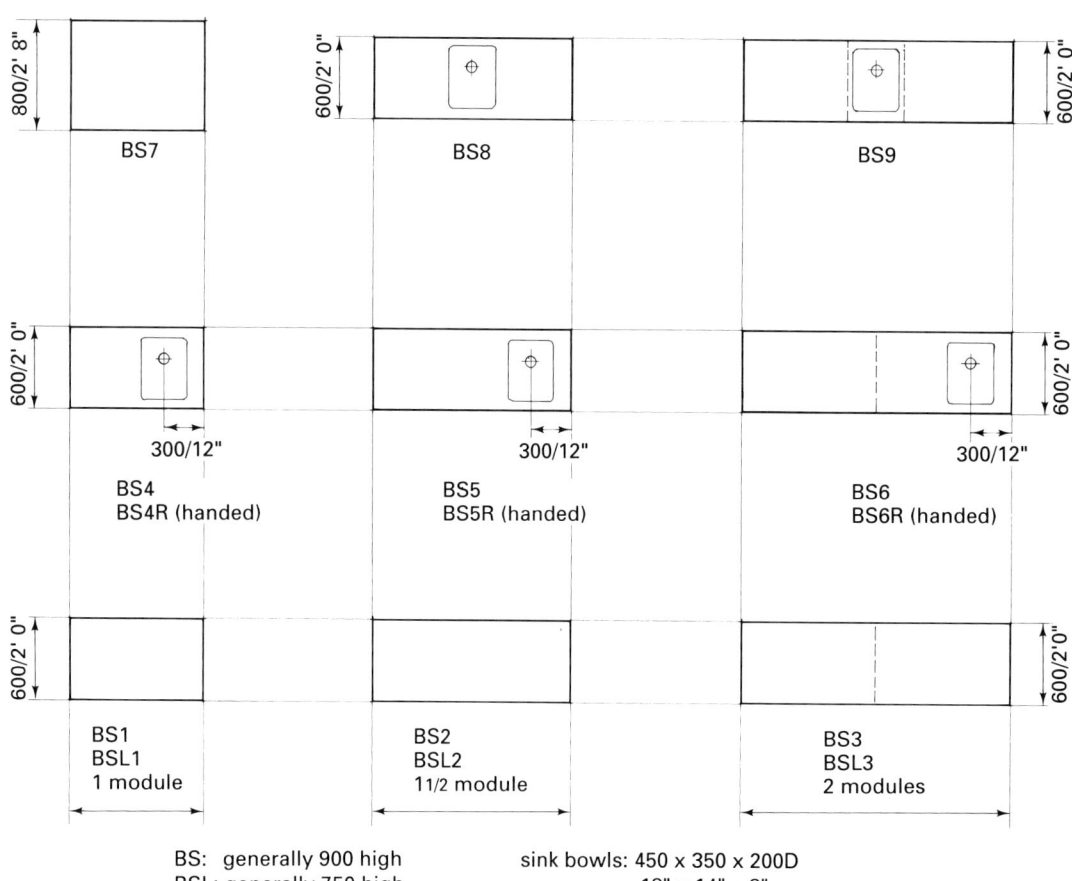

BS: generally 900 high   sink bowls: 450 x 350 x 200D
BSL: generally 750 high   18" x 14" x 8"

Lab furniture: radioactive benches range

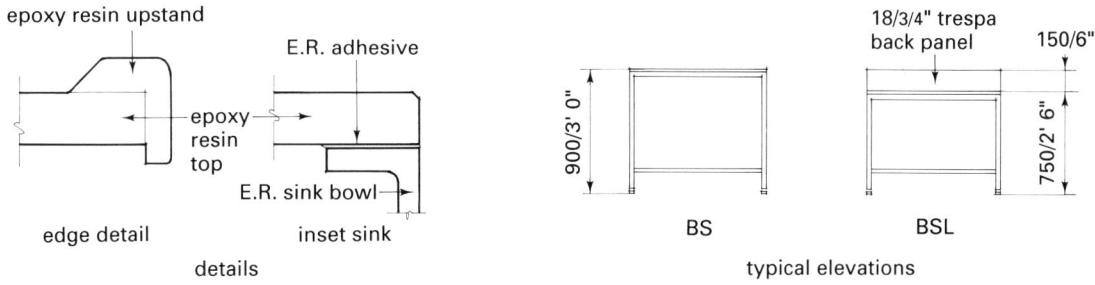

details — typical elevations

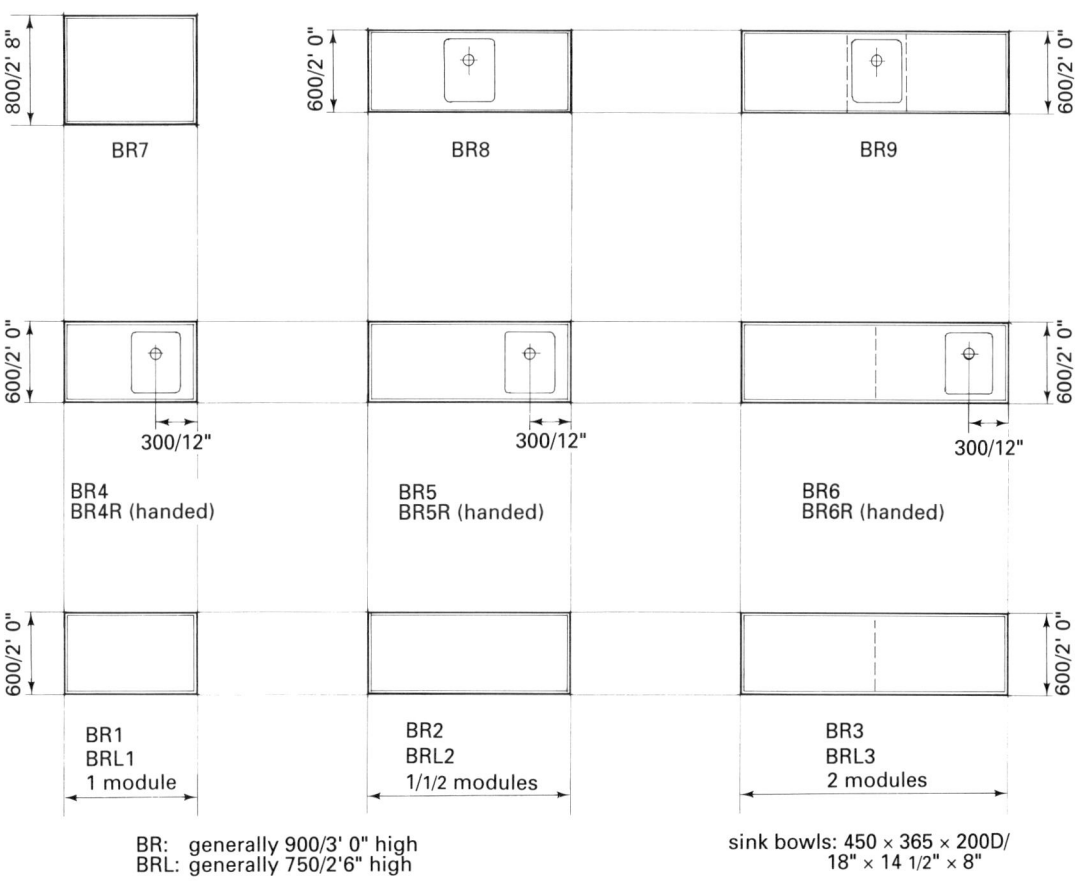

BR: generally 900/3' 0" high
BRL: generally 750/2'6" high

sink bowls: 450 × 365 × 200D/
18" × 14 1/2" × 8"

Laboratory furniture menu

Lab furniture: sink units range

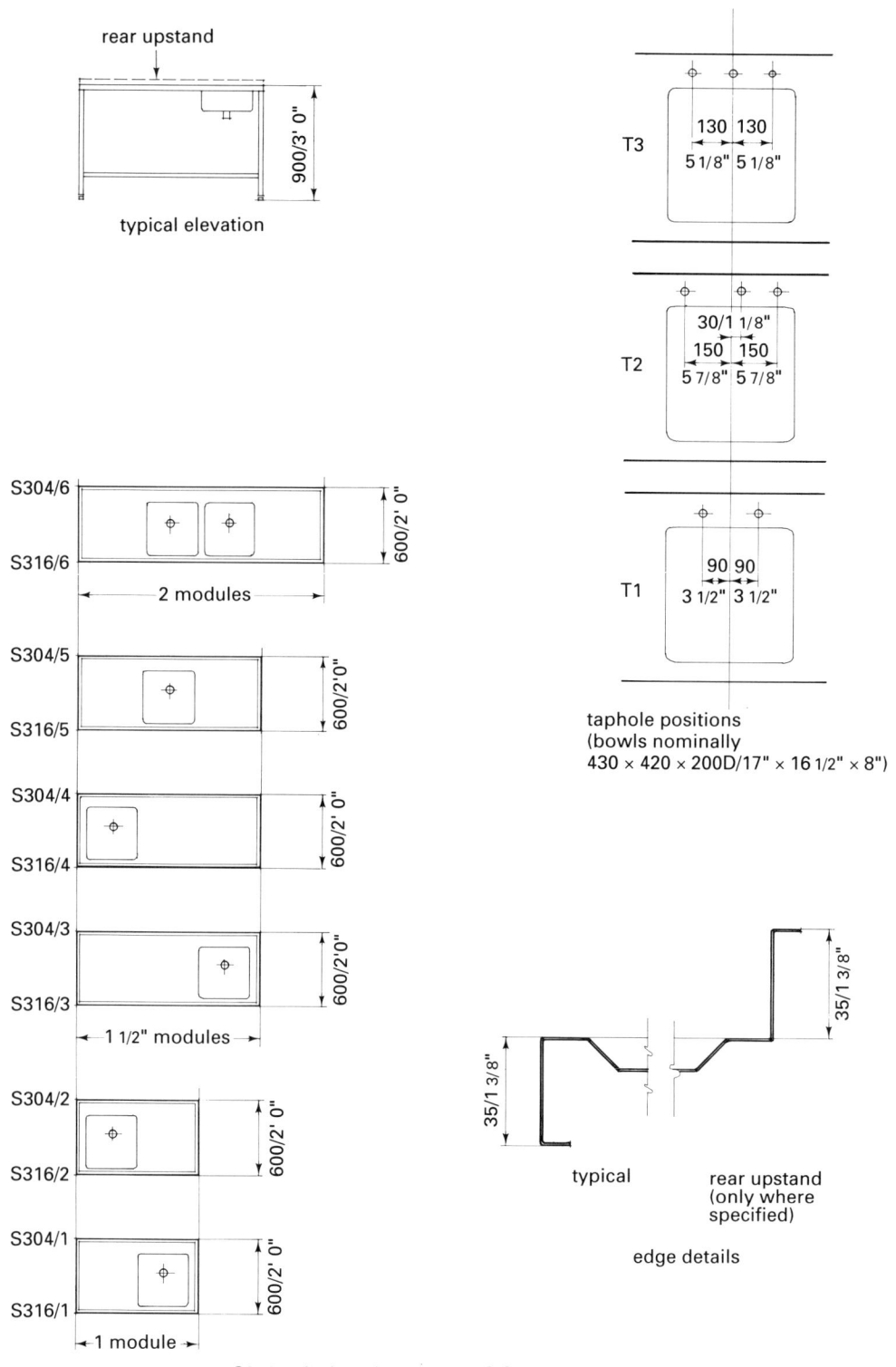

S316 units in 18/8 316 grade S.S. (acid resisting)
S304 units in 18/8 304 grade S.S. (catering)

to be read with sink schedule, where sink code
will be range code + taphole code, e.g. S316/1/T1

Lab furniture: underbench units range

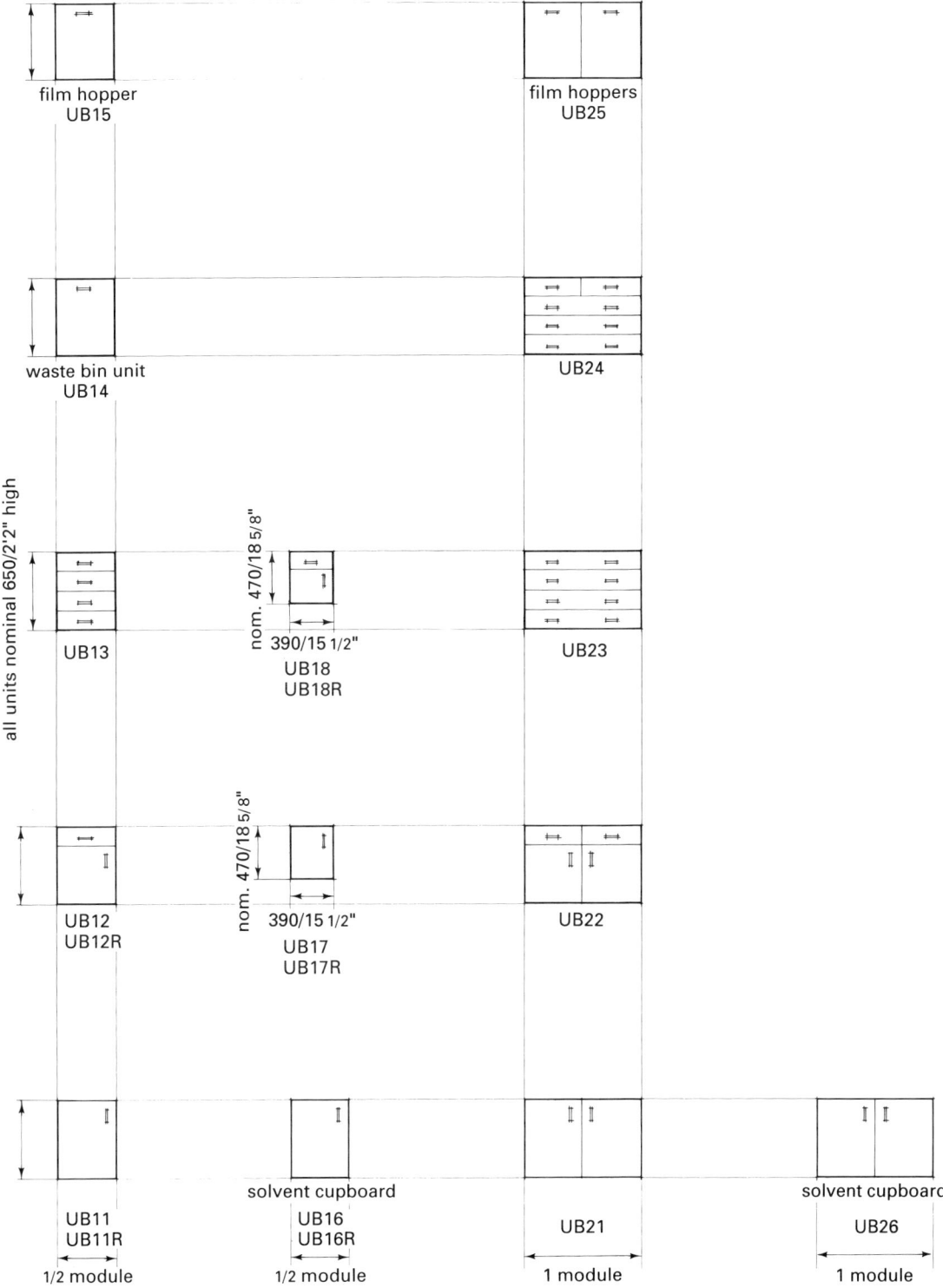

Laboratory furniture menu

Lab furniture: wall cupboards and shelves range

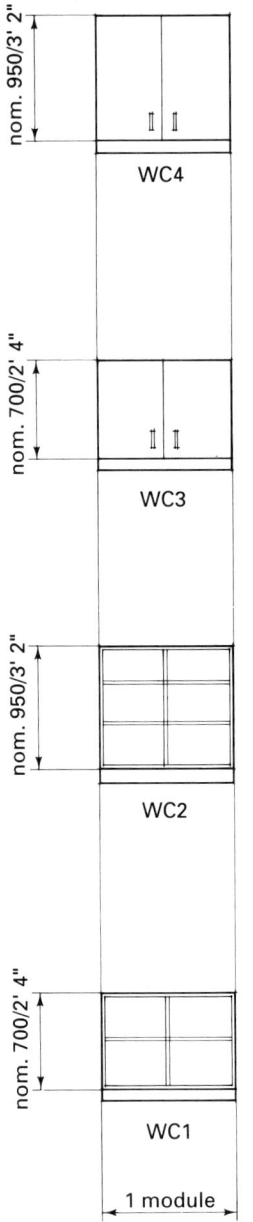

wall cupboards
300 to 350 deep/12" to 14"

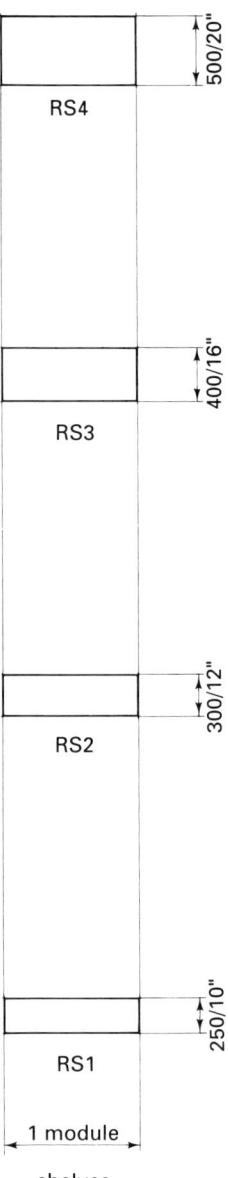

shelves

TS10 Laboratory furniture suppliers

The following table lists a selection of the suppliers in business in the UK when the book was written.

	Supplier	Country of origin	House module (mm)	Pedestal	Bench type Table	Cantilever	Balance tables	Notes
1	Morgan & Grundy Ltd	UK	1000	✓	✓	✓	–	Comprehensive literature. Units in steel or wood
2	ESF	UK	900, 1000	✓	✓	✓	–	
3	Unitform	UK	1000	✓	–	✓	–	
4	Lab Systems Furniture	UK	1000	✓	1200/1400	✓	–	
5	Marson	UK	1000	✓	✓	–	–	
6	Cygnet Joinery Ltd	UK	1000, 1200	✓	✓	✓	–	
7	Isoflow	UK	1000	✓	✓	✓	–	
8	Assab	UK	800	✓	800, 1200 1600	✓	–	Standard items ex-stock. Comprehensive literature. Support post system
9	Lab-Flex Ltd	Denmark	900, 1200	✓	✓	✓	✓	Comprehensive literature. Have worked to l000 mm module on occasion
10	Nordia Lab	Norway	900, 1200	–	✓	✓	–	Comprehensive literature. Support post system. Handled in UK by William Mason & Son
11	S & B	Holland	1200	✓	–	✓	✓	Handled in UK by Hoare Laboratory Engineering
12	Waldner-MSA Ltd	Germany	900, 1200	✓	✓	✓	✓	Comprehensive literature. Support post system
13	Labmarc	USA	1200	–	–	✓	–	Handled in UK by AFOS Ltd

TS11 Fume cupboard criteria

Substance	Relative hazard	Face velocity (m/s)	(ft/min)	Base material	Lining material	Special considerations
Dilute acids, common solvents	**Medium**	0.50	100	**Ceramic**	**Solid laminate**	**Ethers may require flame-proof fan motors**
Concentrated acids; toxic compounds generally	**High**	0.50	100	**Ceramic**	**Varies with substance**	**Benzene may require flame-proof fan motors**
Hydrogen sulphide, nitric acid, sulphur dioxide, hydrochloric acid, hydrofluoric acid	**High**	0.50	100	**Ceramic**	**Stainless steel**	**Hydrofluoric acid requires PVC/ polycarbonate sash. High usage may require fume scrubbing**
Perchloric acid	**High**	0.50	100	**Stainless steel**	**Stainless steel**	**Water wash to cupboard and ducting**
Kjeldahl work	**High**	0.75	150	**Ceramic**	**Stainless steel**	**Frequent cleaning of cupboard required**
Carcinogens, low-energy radioisotopes	**Very high**	0.75	150	**Stainless steel**	**Stainless steel**	**Must be able to be decontaminated**
Violent toxins, high-energy/large quantity radioisotopes	**Extreme**	0.90	220	**Stainless steel**	**Stainless steel**	**Must be able to be decontaminated**

TS12 Fume cupboard schedule

Cupboard Code[a]	Width[b]	Base[c]	Lining[d]	Support[e]	Hazard[f]	Face velocity[g]	Walk in	Extract System Services			Services					
								Water wash	Fume scrubber	13 A SSO	CA	Gas	CW	Sink	Drip cup
FC116/1	1500	SS	SS	UF	H/P	0.50	–	✓	–	4	✓	–	✓	–	✓
FC116/2	2000	ER	PP	UB/V	M	0.50	–	–	–	4	–	✓	2	✓	–
FH206/1	1000	–	–	–	–	–	–	–	–	–	–	–	–	–	–

[a] Code: room numbers are prefixed with FC (fume cupboard) or FH (fume hood), with suffix number showing first, second, etc, in room.

[b] Width: dimension across front of cupboard, in mm (or in).

[c] Base: ceramic (CE), epoxy resin (ER), tiles (TI), solid laminate (SL), stainless steel (SS).

[d] Lining: polypropylene (PP), rigid PVC (RP), epoxy resin (ER), solid laminate (SL), stainless steel (SS).

[e] Support: underbench cupboard unit (UB) or underframe (UF). Suffix V to UB if ventilated.

[f] Hazard: medium (M), high (H), very high (VH), extreme (E); suffix R if radioactive, P if perchloric acid.

[g] Face velocity: in m/s (or f/min).

TS13 Fume cupboard performance specification

1 General

This specification covers fume cupboards and fume extract hoods. It does not cover the fume extract system nor the make-up air supply system.

The documents describing fume cupboards and extract hoods are this Specification, the Schedule of Fume Cupboards and Extract Hoods, and the drawings.

The documents describe the sizes and performance requirements of a range of fume cupboards and extract hoods. Tenderers are free to offer components from their own ranges with their finish and construction wherever these meet the sizes and performance requirements called for in the documents.

Where finishes are specifically called for, such as to bases and linings, these indicate performance requirements and must be complied with.

Tenders must be accompanied by a specification describing the construction and finishes of the components offered.

The fume cupboards shall be supplied complete with dripcups, sinks, electrical services and piped service outlets as shown.

Tenders shall include for the supply to site, assembly, fixing in position and for servicing and adjusting after completion by others of the extract and supply systems and the connection by others of the piped services, waste and power supply.

2 Fume cupboards

2.1 General description of performance requirements

Fume cupboards shall be of the airflow/airfoil type, suitable to accept 70% of the air exhausted by the cupboards from a source independent of the room air.

They shall draw 100% of the air exhausted by the cupboard through the face, the 70% make-up air being supplied to a chamber and directed from the latter down the outside face of the sash, the remaining 30% being air from the room.

They shall have an automatic bypass to direct the make-up air from the chamber direct into the cupboard when the sash is down, to eliminate high face velocities at minimal sash openings.

They shall be fitted with vertical fascias and a horizontal apron of airfoil section, and a pressurized deflector vane under the front apron, to eliminate cross draughts.

They shall operate safely at face velocities of 1.5 m/s with a sash opening of 600 mm.

2.2 General description of construction

Fume cupboards shall be of double-skin construction with the inner skin or liner as specified in the Schedule and the outer skin of zinc coated steel with baked enamel finish, GRP, etc., to approval, and shall be fitted with a removable back baffle of the same material as the liners.

They shall have glass top-lighting panel with reflector and fluorescent lamp complete, switched from front side fascia and wired to connection box at back of cupboard.

They shall be complete with extract outlet and make-up air input of the sizes appropriate to the ventilation requirements for connection to the supply and extract ducting by others.

They shall have a vertically sliding sash of 6 mm frameless laminated safety glass running in PVC side channels, suspended on stainless steel finger grip, complete with fail-safe device in the event of cable breakage.

All bolts and fixings that may come into contact with fumes shall be in stainless steel.

Finishes to bases and liners/baffles shall be as shown in the Schedule.

Junctions between bases and linings, linings and front, and side and back linings, shall be sealed to approval.

2.3 *Sizes*

Cupboards shall be of the widths shown on the Schedule and with depth of approximately 950 mm over the side fascias. Heights from the top of the bases shall be approximately 1200 mm to the top of the cupboard chamber and 1700 mm to the top of the cupboard overall. The height of bases above the floor shall be 900 mm except to walk-in cupboards. The latter shall generally be as above described but shall extend down to floor level with bases at floor level, and shall be fitted with two sashes.

2.4 *Control panel*

Cupboards shall be fitted on one fascia with a panel containing push-button switches and illuminated legends for: Fan ON/OFF, Sash OK, Sash High (audible visual alarm), Fan Fail (audible visual alarm), Extract Fan OK, Input Fan OK, Water Wash Control (where applicable).

2.5 *Services*

Power

Each cupboard shall be fitted with 13 A switched socket outlets as specified in the Schedule in each side fascia, with wiring harness to connection box at back.

Piped services

Cupboards shall be supplied with the services shown on the Schedule, with outlets mounted inside on the side liners, remotely controlled from approved positions on the fascias, and shall be piped back to connection points at the back of the cupboard, with tails to enable the connections to be made to services run-outs by others (run-outs will generally be on back walls below base height).

Cold-water outlets shall be Broen Valves no. 18621 with no. 18719 telescopic rod control valve, or equal approved.

Gas outlets shall be Broen Valves no. 18681 with no. 18716 telescopic rod control valve, or equal approved.

Other outlets shall be Broen Valves no. 18620 with no. 18716 telescopic rod control valve, or equal approved.

Dripcups

Shall be Vulcathene drip wastes with removable strainers and locknuts, 38 mm outlet tails, generally 102 mm circular. Oval dripcups shall be 178 mm.

Sinks

Shall generally be of the same material as the base, complete with Vulcathene 38 mm sink waste with flat locknut.

2.6 *Supports*

Fume cupboards shall be supported on underbench units or underframes as shown in the Schedule.

Underbench units shall be cupboard units with plain rail over, of an overall width to match the fume cupboard, 500 mm deep and approximately 875 mm high to enable the base height to be 900 mm, and shall be complete with plinth and end cover panels where exposed.

Underframes shall be of steel tubing, powder coated, legs with adjustable feet.

Where specified to be ventilated, the cupboard shall be provided with a spigot at the rear for connection by others directly into the extract system.

2.7 Wash-down cupboards

Where specified on the Schedule these shall be provided with internal water wash equipment to approval for periodic washing down of the interior of the cupboard. The equipment shall be remotely controlled from the side fascia by means of Broen no. 18718 telescopic rod control valve, or equal approved.

3   Exhaust hoods

Shall be of the sizes shown on the Schedule, constructed of heavy-gauge aluminium finished with chlorinated rubber paint, for suspending over benches or sinks, complete with internal baffle, outlet spigot for connection to extract duct by others and with provision for fixing to wall at back and suspending from ceiling at front, to approval.

TS14 Schedule of taps and valves

Project: Job no:

	Code	Description	Manufacturer's cat no.
1		Water swan-neck fitting, bench-mounted	
2		Water swan-neck fitting, wall-mounted	
3		Special water fitting, bench-mounted	
4		Special water fitting, wall-mounted	
5		Water fitting with 1 valve, bench-mounted	
6		Water fitting with 2 valves, bench-mounted	
7		Water fitting with 2 valves, wall-mounted	
8		Water fitting with 3 valves, bench-mounted	
9		Water fitting with 3 valves, wall-mounted	
10		Lab H&C mixer fitting, bench-mounted	
11		Lab H&C mixer fitting, wall-mounted	
12		Lab 1 hole H&C mixer fitting, bench-mounted	
13		Inclined sink pillar tap, CP	
14		Sink bib tap, CP	
15		Bib tap CP with wall mount for exposed supplies	
16		Bib tap CP with wall mount for concealed supplies	
17		Standard H&C mixer, bench-mounted	
18		Standard H&C mixer, wall-mounted	
19		1 valve gas cock, bench-mounted	
20		1 valve gas cock, wall-mounted	
21		2 valve gas cock, 90°, bench-mounted	
22		2 valve gas cock, 90°, wall-mounted	
23		2 valve gas cock, 180°, bench-mounted	
24		4 valve gas cock, bench-mounted	
25		Special gases single valve, bench-mounted	
26		Special gases single valve, wall-mounted	
27		Special gases 2 valves 90°, bench-mounted	
28		Special gases 2 valves 90°, wall-mounted	
29		Special gases 2 valves 180°, bench-mounted	
30		Special gases 4 valves, bench-mounted	
31		Oxygen single valve, bench-mounted	
32		Oxygen single valve, wall-mounted	
33		Oxygen 2 valves 90°, bench-mounted	
34		Oxygen 2 valves 90°, wall-mounted	
35		Oxygen 2 valves 180°, bench-mounted	
36		Oxygen 4 valves, bench-mounted	
37		Steam swan-neck, bench-mounted	
38		Steam single valve, bench-mounted	
39		Steam single valve, wall-mounted	

Note: The suffix W will denote wrist/elbow operation to tap.

TS15 Bench outlets: colour code chart

Media	DIN 12 920 recommendations			Text symbols
	Handle	Ring	Button	
Cold water	Green No. 5168	Green	Blue	
Hot water	Green No. 5168	Green	Red	
Demineralized water	Light green No. 5278	Red	Blue	
Low-pressure steam	Red No. 3289	Red	Red	S
Vacuum	Dark Grey No. 2696	Grey	Grey	LV VAC HV
Compressed air	Light blue No. 6188	Grey	Grey	AIR
Oxygen	Light blue No. 6188	Blue	Yellow	OXY
Propane	Yellow No. 4059	Red	Yellow	LIFT GAS TURN GAS PRESS GAS TURN
Acetylene	Yellow No. 4059	Red	White	
Hydrogen	Yellow No. 4059	Red	Red	H
Nitrogen	Yellow No. 4059	Green	Green	N
Carbon dioxide	Yellow No. 4059	Grey	Grey	
Argon	Yellow No. 4059	Grey	Yellow	
Helium	Yellow No. 4059	Grey	White	
Methane	Yellow No. 4059	Red	Blue	
Ethylene	Yellow No. 4059	Red	Black	
Propylene	Yellow No. 4059	Red	Green	
Town gas	Yellow No. 4059	Yellow	Yellow	

Technical supplements

TS16 Services into each laboratory module

1. On side walls to run-outs zone below services spines
   - 1.1 Cold water
   - 1.2 Hot water
   - 1.3 Pressure water – may replace 1
   - 1.4 Purified water – may also go to super-pure unit on wall above sink
   - 1.5 Vacuum, compressed air, special gases
   - 1.6 Waste

2. On side walls above services spines
   - 2.1 Power – to centre compartment of dado trunking
   - 2.2 Telephone – to upper/lower compartment of dado trunking
   - 2.3 Data/computer – to upper/lower compartment of dado trunking
   - 2.4 Special earth – on wall above dado trunking

3. On end wall(s) at low level
   - 3.1 Heating F and R – where central heating

4. On end wall(s) at high level
   - 4.1 General lighting – to lighting trunking where no suspended ceiling, to ceiling void where suspended ceiling
   - 4.2 Emergency lighting – as general lighting
   - 4.3 Smoke detector – to detector on ceiling
   - 4.4 Fume cupboard extract/safety cabinet extract
   - 4.5 General extract – where lab is mechanically ventilated
   - 4.6 Fume cupboard make-up air supply – not required if 4.7
   - 4.7 General fresh air supply – where lab is mechanically ventilated

TS17 A servicing concept for research laboratories

This concept is of particular relevance to research laboratories such as those engaged in medical and pharmacological research, in which the servicing requirements are onerous and alter considerably when a new project is undertaken or when a new team occupies the labs. The concept aims to simplify the initial installation of engineering services, and subsequently to facilitate the addition of new services to individual lab modules or the modification of the existing services, with the minimum of disruption to the module undergoing the changes and also to the occupied space in the rest of the building. It achieves these aims by keeping the principal piped services and electrical submain routes, and all main ductwork routes, outside the occupied space of the building.

The concept does not negate the lab module concept nor any of the other proposals set out in this book for support systems, services spines and bridges, bench services and lab furniture.

The concept places the principal vertical services runs outside the main plan area of the building, contained within a continuous void or duct between the external walls of the lab modules and the external glass envelope to the building, with services branches feeding into each lab module and the void connecting directly to the main plant room.

The concept was the basis of a preliminary proposal made by the author to Cambridge University in 1985 for a laboratory development, which was not proceeded with. In June 1991 the UK *Architectural Review* published details of the University of Alberta's Heritage Medical Research Building in Canada, which uses a similar concept for a multi-storey heavily serviced laboratory building, using the Type 3 plan form described in Chapter 1, section 1.5.2.

The concept can provide the following:

- accommodation for all the vertical drops and risers from the main plant room to the lab modules, together with valved connections to the run-outs in each module and consumer units for each;
- accommodation for extract ducts from fume cupboards in the lab modules to their discharge points above roof level;
- a 'local' plant room for each lab module immediately outside the module;
- in mechanically ventilated buildings, accommodation for supply and extract ducts from the main plant room to each lab module;
- in naturally ventilated buildings, a 'flue' across the face of the building, open at bottom and top, in which the stack effect will induce a vertical airflow into which lab windows can be opened for ventilation;
- an external weatherscreen to the building beyond the external wall to the labs, allowing the latter to become a half-glazed and insulated stud partition;
- access outside the laboratory areas proper for installing, removing and maintaining the services;
- a thermal buffer for the building.

If the void is to be employed as a local plant room then, in addition to housing the shared services drops and risers, it can also house the plant unique to the lab module opposite it, such as fume cupboard extract fan, make-up air-handling unit and condenser unit. Positioning these items of plant locally means reduced space requirements in the main plant room together with reduced cable runs between them and the fume cupboards.

The void can be formed with a light steel-framed structure independent of the building structure but tied back to it for stability, to support the external glass envelope, catwalk, ducts, pipes and plant.

The width of the void will depend upon the extent of the services to be accommodated, the number of floors to be serviced, whether an access catwalk is to be included, and any restrictions on the width of the building imposed by site conditions.

A clear width of about 750 mm (2 ft 6 in) should suffice for a low-rise (two or three stories) building in which the void does not house local plant and in which there is access to the void from each lab module. If local plant is to be housed then this will need to be increased to about 1.5 m (5 ft), as shown in the illustration. In the built example mentioned above at Alberta University, with six floors of labs, an access catwalk, full mechanical ventilation, provision for a fume cupboard in each lab module, submain supply piping and electrical trunking running horizontally beside the catwalk and no site restrictions, the void is 5 m (16 ft 6 in) wide.

The concept is particularly appropriate to the refurbishment of existing buildings (especially if they have not previously been used as labs) provided that the necessary site space exists, as it avoids the need to form large openings in the existing structure and to find space within the existing building for large vertical ducts. The reduced space requirements for the main plant room can also be of assistance.

Servicing concept: plan

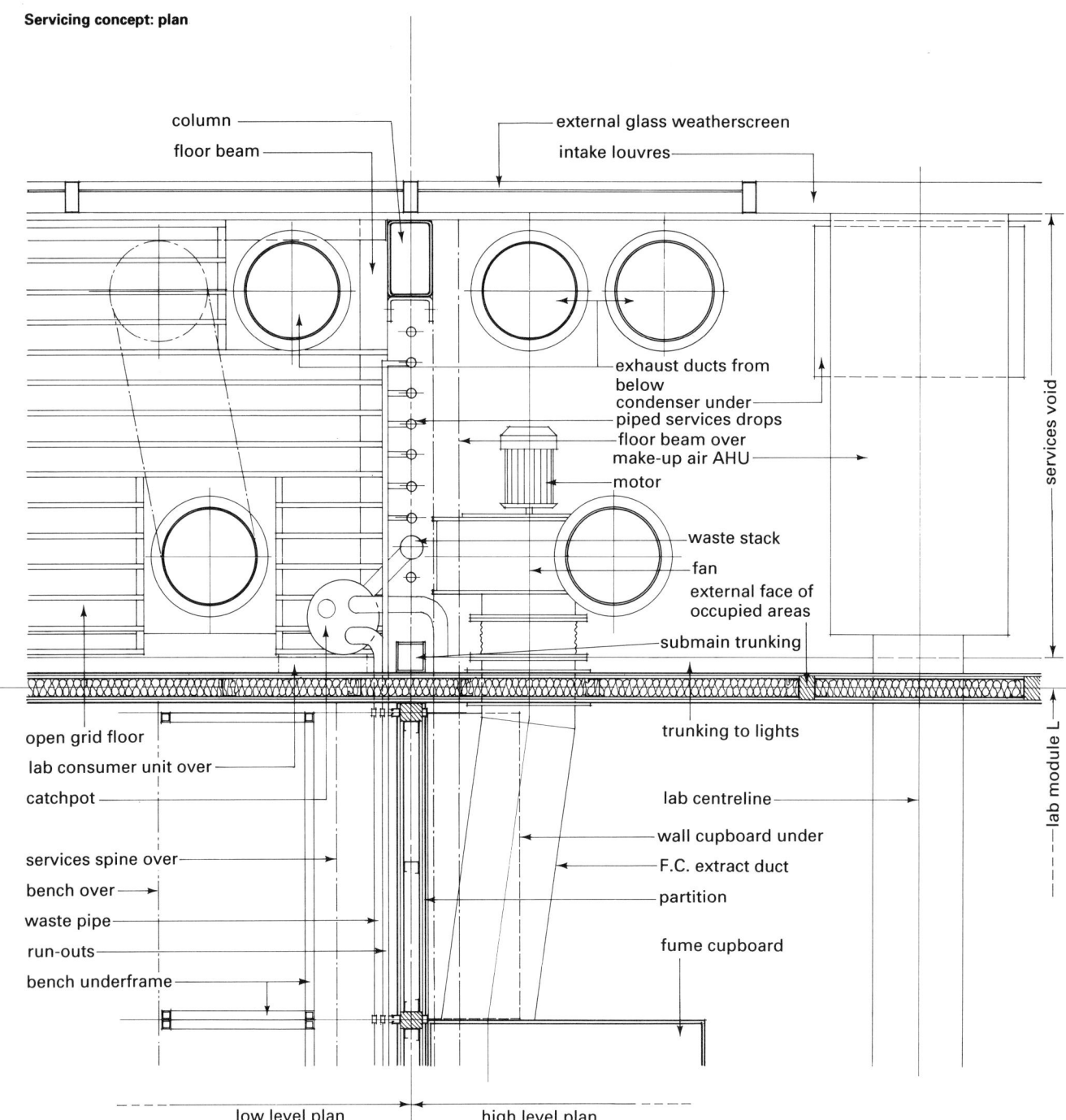

Note: plant supports omitted for clarity

Technical supplements

Servicing concept: section

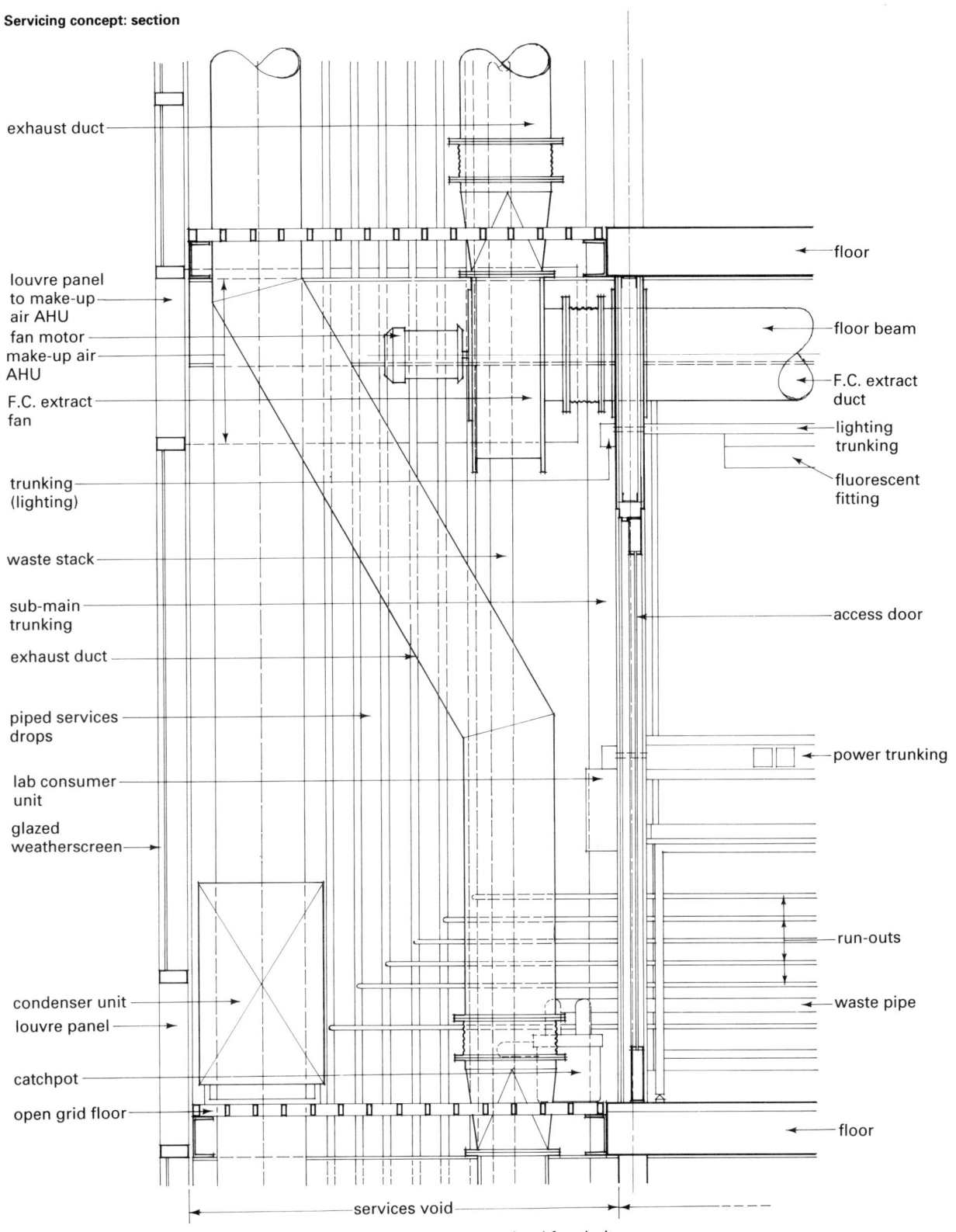

Note: plant supports omitted for clarity

Servicing concept for research laboratories

TS18 Chemical resistance chart

The tabulated information below shows the effect on both polythene and polypropylene of a wide range of chemicals. These results have been obtained from laboratory tests, and when assessing them it should be remembered that unadulterated samples were used. In a typical waste drainage application, however, water and other innocuous fluids would be discharged into the system to have a dilutionary effect on any noxious material that may be present.

The suitability, or otherwise, of a polythene or polypropylene system for use with the chemicals listed, is denoted by A,B,C or D classification.

The letters A or B denote that both polythene and polypropylene may be used without qualification. A chemical with a C classification denotes that care should be taken to ensure adequate dilution of the chemical in question. A chemical with a D classification should only be used after testing.

An upper-case letter represents polypropylene. A lower-case letter represents polythene. A blank means that no information has been published.

Environment	Concentration (%)	Temperature 20 °C	Temperature 60 °C	Environment	Concentration (%)	Temperature 20 °C	Temperature 60 °C
Acetic acid	10 aqueous	Aa	Aa	Barium hydroxide		Aa	Aa
	40	A–	–	Barium sulphate	Saturated	Aa	Aa
	50	A–	A–	Barium sulphide	Saturated	Aa	Aa
	Glacial	Ab	Bc	Beer		Aa	Aa
Acetone	100	Ac	A–	Benzene	100	Bc	Cc
Acetophenone	100	B–	B–	Benzoic acid		Aa	Aa
Acriflavine	2	A–	A–	Benzyl alcohol		Ac	Ac
(2% solution in H$_2$0)				Bismuth carbonate	Saturated	Aa	Aa
Acrylic emulsion		A–	A–	Borax		Aa	Aa
Alums (all types)		Aa	Aa	Boric acid		Aa	Aa
Aluminium chloride		Aa	Aa	Brine	Saturated	Aa	Aa
Aluminium fluoride		Aa	Aa	Bromine liquid	100	Dc	–c
Aluminium sulphate		Aa	–a	Butyl alcohol	100	Aa	–
Ammonia	Aqueous solution 30	Aa	Aa	Calcium carbonate	Saturated	Aa	Aa
Ammonia gas (dry)		Aa	Aa	Calcium chlorate	Saturated	Aa	Aa
Ammonium carbonate	Saturated	Aa	Aa	Calcium chloride	Aqueous solution 50	Aa	Aa
Ammonium chloride	Saturated	Aa	Aa	Calcium hydroxide		Aa	Aa
Ammonium fluoride	20	A–	A–	Calcium hypochlorite	20	Aa	Ba
Ammonium hydroxide	10	Aa	Aa	bleach			
Ammonium	Saturated	Aa	Aa	Calcium nitrate		Aa	Aa
metaphosphate				Calcium phosphate	50	Aa	–a
Ammonium nitrate	Saturated	Aa	Aa	Calcium sulphate		Aa	Aa
Ammonium persulphate	Saturated	Aa	Aa	Calcium sulphite		A–	A–
Ammonium sulphate	Saturated	Aa	Aa	Carbon dioxide	Dry	Aa	Aa
Ammonium sulphide	Saturated	Aa	Aa		Wet	A–	A–
Ammonium thiocyanate	Saturated	Aa	Aa	Carbon disulphide	100	Bc	C–
Amyl acetate	100	Bc	Cc	Carbon monoxide		Aa	Aa
Amyl alcohol	100	Aa	Bc	Carbon tetrachloride	100	Cc	Cc
Aniline	100	Ac	Ac	Carbonic acid		Aa	Aa
Anisole	100	B–	B–	Castor oil		Ac	–c
Antimony chloride		A–	A–	Caustic		Aa	Aa
Aqua regia	Concentrated	Bc	Bc	Cetyl alcohol	100	Ac	–c
Aviation fuel	100	B–	C–	Chlorine	Dry gas 100	Dc	Dc
(115/145 octane)				Chlorobenzene	100	Cc	Cc
Aviation turbine fuel	100	B–	C–	Chloroform	100	Cc	Dc
Barium carbonate	Saturated	Aa	Aa	Chlorosulphonic acid	100	Dc	Dc
Barium chloride	Saturated	Aa	Aa	Chrome alum		Aa	Aa

Environment	Concentration (%)	Temperature 20 °C	Temperature 60 °C
Chromic/sulphuric acid		D–	D–
Cider		Aa	Aa
Citric acid	10	Aa	Aa
Copper chloride	Saturated	Aa	Aa
Copper cyanide	Saturated	Aa	Aa
Copper fluoride	Saturated	Aa	Aa
Copper nitrate	Saturated	Aa	Aa
Copper sulphate	Saturated	Aa	Aa
Cotton seed oil		A–	A–
Cuprous chloride	Saturated	A–	A–
Cyclohexanol	100	Ac	Bc
Cyclohexanone	100	Bc	Cc
Decalin	100	C–	C–
Detergents	2	Aa	Aa
Dibutyl phthalate	100	Ab	Ac
Dichloroethylene	100	A–	–
Diethanolamine	100	A–	A–
Di-iso octylphthalate	100	A–	A–
Emulsifiers	All concentrated	Aa	Aa
Ethyl acetate	100	Bb	Bc
Ethyl chloride	100	Cc	Cc
Ethyl ether	100	B–	–
Ethanaloxime	100	A–	A–
Ethylene dichloride	100	Bc	–c
Ethylene oxide	100	Ba	–
Fatty acids (C$_R$)	100	A–	A–
Ferric chloride	Saturated	Aa	Aa
Ferric nitrate	Saturated	Aa	Aa
Ferric sulphate	Saturated	Aa	Aa
Ferrous chloride	Saturated	Aa	Aa
Ferrous sulphate	Saturated	Aa	Aa
Fluosilicic acid		A–	A–
Fluorosilic acid	40	Aa	Aa
Formaldehyde		Aa	Aa
Formic acid	10	Aa	Aa
	100	Aa	–a
Fructose		Aa	Aa
Fruit juices		A–	A–
Furfural	100	Cc	Cc
Gasoline	100	B–	C–
Gearbox oil	100	A–	B–
Gelatine		A–	A–
Glucose	20	Aa	Aa
Glycerine	100	Aa	Aa
Glycol		Aa	Aa
Hexane	100	A–	B–
Hydrobromic acid	50 aqueous	Aa	Aa
Hydrochloric acid	2	A–	A–
	10 aqueous	Aa	Aa
	20	A–	A–
	30	A–	B–
50 50 HCl HNO$_3$		B–	C–
Hydrocyanic acid	10	–a	–a
Hydrofluoric acid	40	Aa	–a
	60	Aa	Aa
Hydrogen		Aa	Aa
Hydrogen bromide	Dry	Aa	Aa
Hydrogen chloride	Dry 100	Aa	Aa
Hydrogen peroxide	3 (10 vols)	Aa	–a
	10	A–	B–
	30(100 vols)	Aa	Ac
Hydrogen sulphide		Aa	Aa
Hydroquinone		Aa	Aa
Inks	Solution	Aa	Aa
Iodine tincture		A–	–
Isopropyl alcohol	100	A–	A–
Iso-octane	100	C–	C–
Ketones		A–	–
Lactic acid	20	A–	A–
Lanolin	100	A–	A–
Lead acetate	Saturated	Aa	Aa
Linseed oil	100	Ab	Ac
Lubricating oil	100	A–	B–
Magenta dye	2 aqueous	A–	A–
Magnesium carbonate	Saturated	Aa	Aa
Magnesium chloride	Saturated	Aa	Aa
Magnesium hydroxide	Saturated	Aa	Aa
Magnesium nitrate	Saturated	Aa	Aa
Magnesium sulphate	Saturated	Aa	Aa
Magnesium sulphite	Saturated	A–	A–
Manganese sulphate		Aa	Aa
Meat juices		A–	A–
Mercuric chloride	40	Aa	Aa
Mercuric cyanide	Saturated	Aa	Aa
Mercury	100	Aa	Aa
Mercurous nitrate	Saturated	Aa	Aa
Methyl alcohol	6 aqueous	–a	–
	100	Ab	Ac
Methyl ethyl ketone		Cb	–c
Methylene chloride	100	Ac	Ac
Milk and its products		Aa	B a
Mineral oils		Ab	Ac
Molasses	100	Aa	Ba
Motor oil	100	A–	A–
Naphthalene	100	Ac	Ac
Nickel chloride	Saturated	Aa	Aa
Nickel nitrate	Saturated	Aa	Aa
Nickel sulphate	Saturated	Aa	Aa
Nitric acid	10	Aa	Aa
	60	B–	D–
	70	Cb	Dc
50 50 HNO$_3$ HCl		B–	D–
50 50 HNO$_3$ H$_2$SO$_4$		C–	D–
Nitrobenzene	100	Ac	Ac
Oleic acid		Aa	B–
Oleum		Dc	Dc
Olive oil	100	A–	A–
Oxalic acid	50 aqueous	Aa	Ba
Paraffin	100	Ab	Bc
Paraffin wax	100	A–	A–
Petrol	100	Bc	Cc
Phenol	100	Ac	Ac
Phosphoric acid	95	Ab	Ac
Photographic developers		Aa	Aa

Chemical resistance chart

Environment	Concentration (%)	Temperature 20 °C	60 °C
Photographic emulsions		Aa	A–
Plating solutions: brass		A–	A–
Plating solutions: cadmium		A–	A–
Plating solutions: chromium		A–	A–
Plating solutions: copper		A–	A–
Plating solutions: gold		A–	A–
Plating solutions: indium		A–	A–
Plating solutions: lead		A–	A–
Plating solutions: nickel		A–	A–
Potassium bicarbonate	Saturated	Aa	Aa
Potassium fluoride		Aa	Aa
Potassium hydroxide		Aa	Aa
Potassium nitrate	Saturated	Aa	Aa
Potassium perborate	Saturated	Aa	Aa
Potassium perchlorate	10	Aa	Aa
Potassium permanganate	20	Aa	Aa
Pyridine	100	A–	–
Silicone oil	100	A–	A–
Soap solution	Concentrated	Aa	Aa
Sodium acetate	Saturated solution	Aa	Aa
Sodium bisulphate	Saturated	Aa	Aa
Sodium bisulphite	Saturated	Aa	Aa
Sodium borate		Aa	Aa
Sodium cyanide	Saturated	Aa	Aa
Sodium hypochlorite	15 chlorine	Aa	Aa
Stannic chloride	Saturated	Aa	Aa
Stannous chloride	Saturated	Aa	Aa
Sulphuric acid	10 aqueous	Aa	Aa
	50	Aa	Aa
	60	Aa	Aa
Sulphuric acid		A–	A–
50 50 H_2SO_4/HNO_3		C–	D–
Tartaric acid	10 aqueous	Aa	Aa
Tetrahydrofuran	100	Cc	Cc
Tetralin	100	C–	C
Toluene	100	Cc	Cc
Transformer oil	100	Ab	Cc
Trichloracetic acid	100	A–	A–
Trichlorethylene	100	Cc	Cc
Triethanolamine	100	Ab	Ac
Vinegar		Aa	Aa
Whiskey		Aa	A–
White paraffin	100	A–	B–
White spirit	100	B–	C–
Wines and spirits		Aa	A–
Xylene	100	Cc	Cc
Yeast		Aa	A–
Zinc chloride	Saturated	Aa	Aa
Zinc oxide		Aa	Aa
Zinc sulphate	Saturated	Aa	Aa

TS19 Radioactive shielding

Because of its density, lead has traditionally been the protective material used (as it is still in doors, screens and aprons) and protective requirements are always related to it, being expressed as 'equivalent to x millimetres of lead'.

In the UK the authority responsible for deciding the protective requirement for a given level of radioactive emission is the National Radiological Protection Board.

Once the required lead thickness is known, the required thickness of the materials that are to be used for the walls/floor/door/observation window to protect those working in adjoining rooms can be calculated. This is done by relating the density of the material to that of lead, i.e. a material with a density one quarter that of lead will require a thickness four times that of the required lead thickness. The table below gives equivalent thicknesses for a range of common materials.

Radioactive emissions in laboratories will normally be from radioisotopes. The measure of radioactive emission is the curie (Ci), which is 2.2×10^{12} nuclear transformations (changes) per minute. Most emissions in labs are only a fraction of this and are expressed in millicuries (one thousandth of a curie), symbol mCi. (Note that the curie will gradually be replaced by the becquerel (Bq), which is the SI-preferred unit. 1 Ci = 3.7×10^{10} Bq, so 1 mCi = 3.7×10^{7} Bq.)

In a typical Grade B laboratory the level of emission that is present at any one time is likely to be 1 mCi, although a maximum of 10 mCi is a useful figure to allow for in the absence of definite information, which should always be sought from the client

Material thickness equivalents (rounded up)

Material	Density (kg/m³)	(lb/ft³)	Thickness (mm)	(in)
Lead	11 340	710	1.0	0.04
Copper	9 000	562	1.3	0.05
Steel	7 850	490	1.5	0.06
Aluminium	2 770	173	4.0	0.16
Glass	2 560	160	4.5	0.18
Lead glass	4 800	300	2.4	0.08
Reinforced concrete	2 500	156	4.5	0.18
Brickwork	1 700	106	6.7	0.27
Blockwork	760	48	15.0	0.60
Timber	380	24	30.0	1.20
	900	56	12.6	0.51
Dense plaster	1 440	90	8.0	0.32
Lightweight plaster	640	40	17.8	0.71

TS20 Fire extinguishers

	Class	Extinguisher type
A	Fires in wood, paper, straw, textiles, other combustibles containing carbonaceous material	Water, foam, spray, powder
B	Fires involving petrol, oils, fats, solvents, paints and other flammable liquids	Powder, foam, spray, CO_2 in early stages
C	Fires involving flammable gases: methane, acetylene, manufactured or natural gas and other flammable gas	Powder, spray for smaller fires
D	Fires involving combustible metals	Powder from low-velocity applicators
E	Fires involving electrical equipment	CO_2, powder, spray

TS21 Scientific disciplines

(From *The Architects' Journal* Information Library 6 January 1965)

The structure of science is traditionally divided into three main disciplines: chemistry, physics and biology.

1   Chemistry

Chemistry is the science of the elements and their compounds and their laws of combination and behaviour in various conditions. It is safe to assume that precautions are always necessary against chemical attack and spillage. Physical chemistry, with its specialized branches of electrochemistry and thermochemistry, forms a series of activities between chemistry and physics but is not regarded as a major discipline because it deals in chemical substances.

1.1 Organic chemistry

The chemistry of substances having carbon in their molecules (formerly applied only to compounds obtained from living organisms, their products or remains). It includes *biochemistry*, the study of the chemical and physicochemical processes and products of the life phenomena of plants and animals (broadly included in organic chemistry but with greater stress on the continuing life process), and *histochemistry*, the study of chemical distribution in tissues. Organic chemistry makes major use of organic solvents: oils, paraffins, alcohols, fatty acids, ketones, ethers, carbohydrates, benzene hydrocarbons, phenols and the heterocyclics such as pyridine.

1.2 Inorganic chemistry

The chemistry of substances other than those included under organic. It generally concerns substances of mineral origin and includes many specialized branches such as metallurgy, mineralogy, geology and cement chemistry. Each of these however has organic connotations: for example, organic hydrocarbons are obtained from mineral oils. Inorganic chemistry makes use of the whole range of inorganic compounds, acidic and alkaline.

2   Physics

Physics is the science concerned with treating the properties of matter and energy, or with the action of different forms of energy on matter in general.

It usually consists of chemically clean processes (except in physical chemistry) and usually demands physical precautions taken such as those against vibration, electrical interference, ionizing radiation, and dust exclusion.

Traditional divisions of the subject are properties of matter, heat, light, sound, electricity, magnetism, but divisions have become successively blurred since the late 19th century by kinetic theory, the theory of relativity, quantum theory (wave mechanics), atomic and nuclear theories.

Application of physical theory has resulted in technologies that have applications in every branch of laboratory activity, such as electronics, radiology and photoelectricity, and their numerous derived techniques involving all degrees of instrumentation from simple electrical measuring techniques to computers, radio and telecommunications, and electron microscopy.

Other subject or discipline names that have laboratory design relevance and usually involve application of physical theories include astronomy, mathematics, statistics, biometrics, bionomics, engineering (mechanical, structural, electrical), biophysics and medical physics (the application of physics to the study of living things: usually implies sophisticated electronic and radioactive investigation techniques). Mathematics, statistics, biometrics and bionomics generally involve the use of computers.

3   Biology

Biology is the science of physical life, dealing with the morphology, physiology, origin and distribution of animals and plants.

The proliferation of specialities in this field is great, owing to the fact that the medical sciences are included in it, although most of the human medical sciences are equally applicable to other animals and indeed the fundamental studies are conducted exclusively on animals rather than on humans. Many activities are carried out in normal laboratory conditions but some biological techniques demand closely controlled environmental conditions: e.g. sterile air supply, temperature controls down to ± 0.25 °C (± 0.5 °F), dust control down to specified particle size, and humidity control.

The traditional division of biology into zoology and botany is still maintained in school curricula and in the nomenclature of university departments. The names therefore crop up regularly in relation to laboratory projects. It must be realized however that in the study of plants and animals (whether human or otherwise) the main subdivisions apply: i.e. the study of normal physical processes (physiology), the study of diseases of the living organism (pathology) and the subjects related to curative, corrective treatments (pharmacology, therapeutics). There are many subdivisions of these three and the distinction between them is often not clear cut. It will be obvious also that biochemistry is a sort of servant discipline which workers in any of the biological disciplines may wish to use, just as many of the physical techniques (electron microscopy, for example) were specially developed to serve the biological disciplines. The classification below is therefore one of convenience: the overlapping between subjects is inherent and many techniques are used by all of them.

Zoology

Zoology is concerned with the natural history of animals, science of their structure, physiology (see below), habits, classification and distribution. It includes studies of specialized groups of animals, such as entomology (insects) and helminthology (worms).

Physiology

Physiology is the study of the normal life processes of humans, animals or plants (plant physiology).

Anatomy

Anatomy is the study of physical structure (usually only of humans or animals). There are many specialized subdivisions of anatomy, including haematology (blood), histology (tissue), and embryology. There are also the broader-based subjects on both animal and plant sides: anthropology (range and origins), ecology (relationships and environment), pathology (the study of diseases in humans or animals). Subdivisions in this category have become major disciplines in themselves, such as bacteriology and virology.

Pharmacology

Pharmacology was originally the study of pharmaceuticals but now extends through many branches of the curative, therapeutic aspects of biology. There are many applied subjects in this field in which a great deal of conventional laboratory work is carried out: medicine, surgery and other specialities which come under these two blanket terms.

Botany

Botany is the science of plants, their structure and classification. Subdivisions include genecology (population and habitat), phenology (periodicity and climate), phytopathology or plant pathology (study of diseases in plants), and parasitology. There are various subjects that cover specialist but widely applicable aspects of biology, such as cytology (cells), genetics (heredity and variation), cytogenetics, and the widely applied term microbiology, which serves to cover the whole field of study surrounding the microscopic organization of living matter.

TS22 External design temperatures
From the IHVE Guide

Country	Town	Winter (°C)	(°F)	Summer (°C)	(°F)
Algeria	Algiers	3	37.4	37	98.6
Argentina	Buenos Aires	−1	30	36	97
Australia	Canberra	−1	30	36	97
	Sydney	7	44.6	35	95
Austria	Vienna	−15	5	31	88
Belgium	Brussels	−10	14	31	88
Brazil	Rio de Janeiro	13	55.4	34	93
Canada	Montreal	−23	−9.4	32	89.6
	Ottawa	−27	−16.6	33	91
Cuba	Havana	15	59	33	91
Czech Republic	Prague	−16	3.2	32	89.6
Denmark	Copenhagen	−12	10.4	28	82
Ethiopia	Addis Ababa	1	33.8	27	81
Finland	Helsinki	−27	−16.6	27	81
France	Paris	−5	23	32	89.6
Germany	Berlin	−15	5	32	89.6
Greece	Athens	−1	30	37	98.6
Hong Kong	Hong Kong	7	44.6	33	91
Hungary	Budapest	−15	5	34	93
India	New Delhi	4	39.2	44	111
Iran	Tehran	−4	24.8	40	104
Iraq	Baghdad	2	36	47	116.6
Italy	Rome	−1	30	36	97
Japan	Tokyo	−2	28.4	33	91
Kuwait	Kuwait City	4	39.2	45	113
Lebanon	Beirut	4	39.2	33	91
Netherlands	Amsterdam	−7	19.4	28	82.4
New Zealand	Christchurch	−1	30	36	97
	Wellington	4	39.2	26	79
Norway	Oslo	−17	1.4	27	81
Philippines	Manila	17	62.6	36	97
Poland	Warsaw	−20	−4	32	89.6
Portugal	Lisbon	3	38	34	93
Romania	Bucharest	−20	−4	36	97
Spain	Madrid	−4	24.8	36	97
Sweden	Stockholm	−19	−2.2	27	81

Country	Town	Winter (°C)	(°F)	Summer (°C)	(°F)
Switzerland	Berne	−12	10.4	30	86
Turkey	Ankara	−14	6.8	36	97
UK	London	−1	30	28	82
USA	Chicago	−20	−4	34	93
	New York	−12	10.4	36	97
	San Francisco	2	36.6	29	84
Russia	Moscow	−26	−15.2	31	88
Yugoslavia	Belgrade	−13	8.6	37	98.6

Warm weather temperatures

Country	Town	(°C)	(°F)
Afghanistan	Kabul	37	98.6
Angola	Luanda	32	89.6
Bangladesh	Dhaka	36	97
Bolivia	La Paz	23	73.4
Botswana	Ghanzi/Maun	37	98.6
Cameroon	Douala	32	89.6
Chad	Ndjamena	43	109.4
Chile	Santiago	33	91.4
China	Shanghai	37	98.6
	Tianjin	38	100.4
Colombia	Bogotá	23	73.4
Congo	Brazzaville	35	95
Ecuador	Quito	27	80.6
Egypt (UAR)	Alexandria	37	98.6
	Cairo	41	106
Ghana	Accra	33	91
Guatemala	Guatemala City	31	88
Indonesia	Jakarta	33	91
Israel	Jerusalem	36	97
Ivory Coast	Abidjan	33	91
Jamaica	Kingston	34	93
Jordan	Amman	38	100.4
Kenya	Mombasa	33	91
	Nairobi	28	82
Korea, South	Seoul	35	95
Liberia	Monrovia	32	89.6
Libya	Tripoli	39	102

Technical supplements

Country	Town	(°C)	(°F)
Madagascar	Antananarivo	31	88
Malawi	Zomba	34	93
Malaysia	Kuala Lumpur	35	95
Mali	Gao	44	111
Mexico	Mexico City	28	82.4
Morocco	Casablanca	33	91.4
Mozambique	Maputo	38	100.4
Myanmar	Yangon	39	102
Namibia	Windhoek	33	91
Nigeria	Ibadan	37	98.6
	Lagos	34	93
Pakistan	Karachi	39	102
Yemen	Aden	39	102
Saudi Arabia	Riyadh	44	111
Senegal	Dakar	37	98.4
Somalia	Mogadishu	34	93
South Africa	Cape Town	34	93
	Johannesburg	31	88
Sri Lanka	Colombo	33	91
Sudan	Khartoum	45	113
Syria	Damascus	41	106
Tanzania	Dar es Salaam	33	91.4
Thailand	Bangkok	38	100.4
Tunisia	Tunis	42	107.6
Uganda	Kampala	32	89.6
Vietnam	Hanoi	38	100.4
	Ho Chi Minh City	37	98.6
Western Sahara	Ad Dakhla	35	95
Zaire	Kalémié	32	89.6
	Kinshasa	35	95
Zambia	Lusaka	34	93
Zimbabwe	Harare	32	89.6

TS23 Conversion data

Length	1 inch (in)	= 25.4 mm or 2.54 cm
	1 foot (ft)	= 304.8 mm or 30.48 cm
	3 ft 3.37 in	= 1 metre or 1000 mm or 100 cm
	1 yard	= 0.914 m or 914 mm or 91.4 cm
	1 mile (mi)	= 1609 m or 1.609 km
Area	1 in^2	= 645 mm^2 or 6.45 cm^2
	1 ft^2	= 0.093 m^2 or 929 cm^2
	10.76 ft^2	= 1 m^2
	1 yd^2	= 0.836 m^2
	1 mi^2	= 2.59 km^2 or 259 hectare (ha)
	1 acre	= 0.4047 ha or 4046.9 m^2
	2.471 acres	= 1 ha or 10 000 m^2
Volume	1 in^3	= 16 387 mm^3 or 16.387 cm^3
	1 ft^3	= 0.028 m^3
	1 yd^3	= 0.7646 m^3
	35.33 ft^3	= 1 m^3
Mass	1 lb	= 0.45 kg
	2.2 lb	= 1 kg
	2200 lb	= 1 tonne (t) or 1000 kg
	1 UK ton	= 1.016 t
	1 US ton	= 0.907 t
Capacity	1 UK pint	= 0.5683 litres (l)
	1 UK gallon	= 4.546 l
	1 US liquid pint	= 0.4732 l
	1 US gallon	= 3.785 l
Mass per unit area	1 lb/in^2	= 703.07 kg/m^2
	1 lb/ft^2	= 4.882 kg/m^2
Mass per unit volume	1 lb/in^3	= 27680 kg/m^3
	1 lb/ft^3	= 16.02 kg/m^3
Temperature	X °Fahrenheit (F)	= $5/9 \times (X-32)$ °Celsius (C)
	X °C	= $(X \times 9/5 + 32)$ °F
	1 °F	= 0.5556 °C

Further reading

The Design of Research Laboratories

Nuffield Foundation, Oxford University Press, London, 1961

The report of a study carried out by the Division for Architectural Studies of the Nuffield Foundation, which established the guidelines for lab design in the UK.

Laboratories, Design, Safety and Project Management

Editor J. Komoly, Ellis Horwood, London, 1992

Intended principally for the lab manager and for people working in labs; not primarily a book for lab designers.

Guidelines for Laboratory Design: Health and Safety Considerations

Di Berardinis, Baum, First, Gatwood, Groden and Seth, Wiley Interscience, 2nd edn, 1993 (£41.95)

Tailored to the USA, not primarily for the lab designer but rather the definitive book on labs. Contains information useful to lab designers but also contains a wealth of information not relevant to their requirements. Useful further reading for the experienced designer.

Laboratory Animal Houses – A Guide to the Design and Planning of Animal Facilities

G. Clough and M.R. Gamble, Medical Research Council Laboratory Animals Centre, Carshalton, Surrey, UK, 1976

Contains much information that is of use to the animal house designer.

Various papers of the Laboratories Investigation Unit, sponsored by the Department of Education and Science and the University Grants Committee.

Index

Acetylene 65
Air conditioning 77
Ancillary spaces 22
Animals
  accommodation 32
  cages 58
Architect 85
Autoclave 55

Benches
  benchtop
    materials 43
    general 42
    menu 102, 103
    systems 44
Bench services 61
Biochemistry lab 14
Biohazardous lab
  criteria 89
  description 16
  example 15
  regulations 89
Biological safety cabinets 91
Brief
  first stage 6
  general 6
  second stage 7
Builder's work ducts 83, 84
Builder's work in connection 82
Building services engineer 85

Catchpot 65, 67
Ceilings 82
Centrifuge 53
Chemical store 32
Chemistry lab 14
Clean rooms
  description 25
  example 26
  regulations 92
Clocks 70
Cold room/cold store
  description 23
  example 24
Common room 27
Compressed air 64

Computer
  room 27
  systems 70
  trunking 73
Constant temperature room, *see* Hot room
Controls 77
Cooling 76
Copier room 28
Corridor width 35
Cylinder store 31

Dangerous pathogens, *see* Biohazardous lab
Darkroom
  description 23
  example 24
Data systems 70
Deep freezer 53
Deionization 64
Design concept 7
Design team 85
Dilution recovery trap, *see* Catchpot
Dishwasher 55
Distillation 64
Doors 81
Drains 68
Drenchshower 71, 72
Dripcup 65
Dryer 55
Drying oven 55

Earth 70
Electron microscope suite
  description 25
  example 26
  instrument 55
Electronic workshop 32
Electrophysiology lab 14
Elevator, *see* Lift
Emergency shower, *see* Drenchshower
Equipment
  categories 53
  guide 54
Equipment room
  description 22
  example 24

External design temperatures 127
Eyewash 71, 72

Fire
  alarm systems 72
  extinguishers 58, 124
  protection systems 72
Fire compartment penetrations 84
Floor data plan
  example 8
  use of 7
Floor finishes 81
Freeze dryer 55
Fuel gas 64
Fume cupboard
  criteria 108
  general 49
  performance spec 110
  schedule 109
  ventilation 74, 75
Fume hood 52

Gas, *see* Fuel gas
Genetic manipulation lab
  criteria 90
  description 16
Glass washer 55

Heating 76
Hot room
  description 23
  example 24
Hydrogenation unit 27

Ice-making machine 55
Incubator 55
Initial questionnaire 94
Inset sinks 49
Instrument room 23
Instrument workshop 32

Lab furniture
  general 42
  menu 100
  performance spec 96
  suppliers 107

Lab gases 64
Lab module
　criteria 21
　general 20
Lab services 62
Lab taps 61, 63, 113
Laminar-flow
cabinet 55
Library 28
Lifts 35
Lighting 70
Louvres 82

Magnetic resonance
imager
　equipment 58
　example 26
　room 25
Mains supplies
　electricity 61
　fuel gas 61
　water 61
Medical research lab
areas 19
Microbiological safety
cabinets
　classifications 91
　description 55
Modular planning 9
Module
　lab 20
　lab furniture 42

Non-standard labs 13

Offices 27

Partitions
　finishes 79
　systems 79
Perchloric acid 51
Piped services
　bench outlets 61
　distribution systems
　60, 61
　materials 65
　types 64
Plan forms 9, 11
Planning
　general 12
　modular 9

Plant room
　flooring 81
　layout 31
　planning 28
　position 28
　size 28
Power
　non-standard 68
　small power 68
　supply 68
　trunking 69
Pressure gas control
console 65
Programmes 86

Quantity surveyor 85

Radioactive lab
　classifications 88
　description 13
　example 15
Radioactive
shielding 123
Radioactive store 31
Refrigerators 53
Regulations 5
Research labs 3, 4
Reverse osmosis 64
Room data sheet
　example 95
　use of 7
Routine labs 3

Safeguarding work
and specimens 74
Safety
　chemicals 72
　fire 72
Safety cabinets,
see Microbiological
safety cabinets
Scientific
disciplines 125
Scintillation
counter 53
Security 72
Seminar room 27
Services
　bridges 40
　into lab module 115
　spines 40

Shelves
　general 47
　menu 106
Sinks
　general 47
　menu 104
　units 49
Solvent store 31
Space standards 17,
18, 20
Special gases,
see Lab gases
Specialist labs 13
Standard labs 13
Steam 65
Stills 55
Storage areas 31
Structural bay 22
Structural
engineer 85
Support systems 36, 37
Standards 5

Teaching labs 3, 4
Telephones 70
Tissue culture lab
　description 16
　example 15
Tundish, see Dripcup

Underbench units
　general 45
　menu 105

Vacuum 64
Ventilation 74
Vertical ducts 10

Wall bar systems 47
Wall cupboards
　general 47
　menu 106
Wash-up room
　description 25
　example 26
Wastes 65
Water
　pressure 64
　purified 64
　services 64
Workshops 32